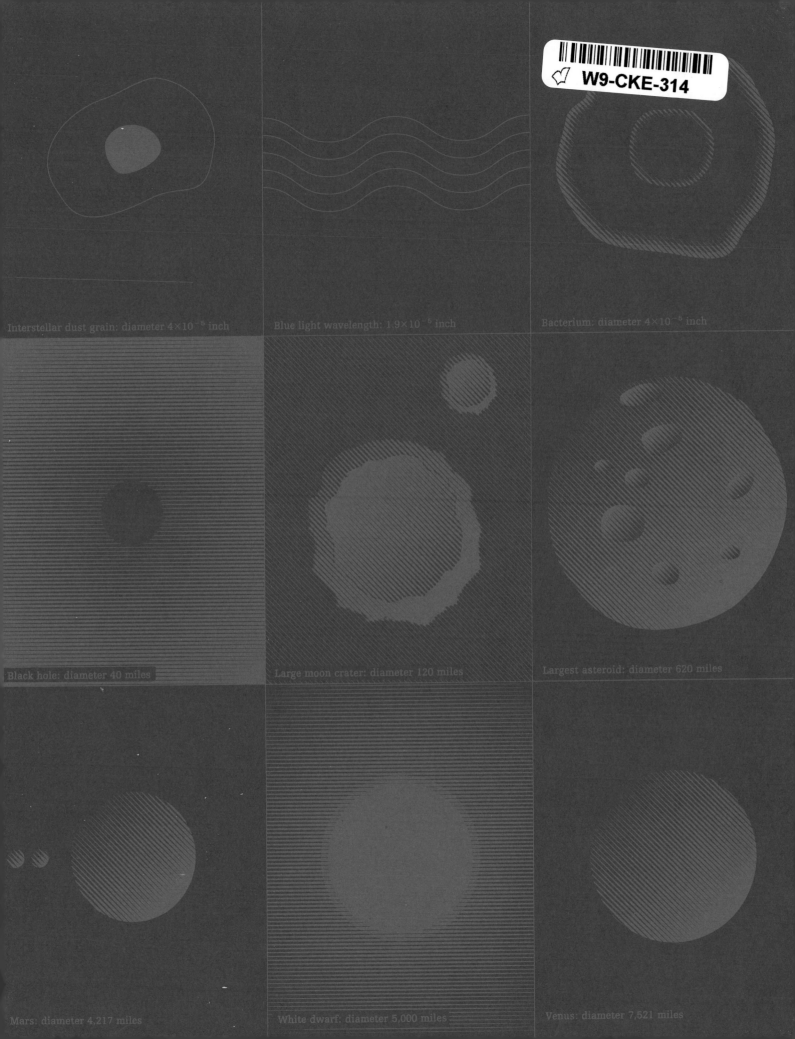

Interstellar dust grain: diameter 4×10⁻⁵ inch

Blue light wavelength: 1.9×10⁻⁵ inch

Bacterium: diameter 4×10⁻⁵ inch

Black hole: diameter 40 miles

Large moon crater: diameter 120 miles

Largest asteroid: diameter 620 miles

Mars: diameter 4,217 miles

White dwarf: diameter 5,000 miles

Venus: diameter 7,521 miles

# OUTBOUND

```
Q                    483298
629.4                21.27
Out                  Nov89
Outbound
```

```
Q                    483298
629.4                21.27
Out                  Nov89
Outbound
```

| DATE | ISSUED TO |
|------|-----------|
|      |           |
|      |           |
|      |           |
|      |           |

The Moon floats above the blue glow of the terrestrial atmosphere, beacon and lure for the first interworld explorers from Earth.

Dropping away toward the Moon's cratered surface, the lunar module *Intrepid* ferries the landing crew of *Apollo 12* to the Ocean of Storms.

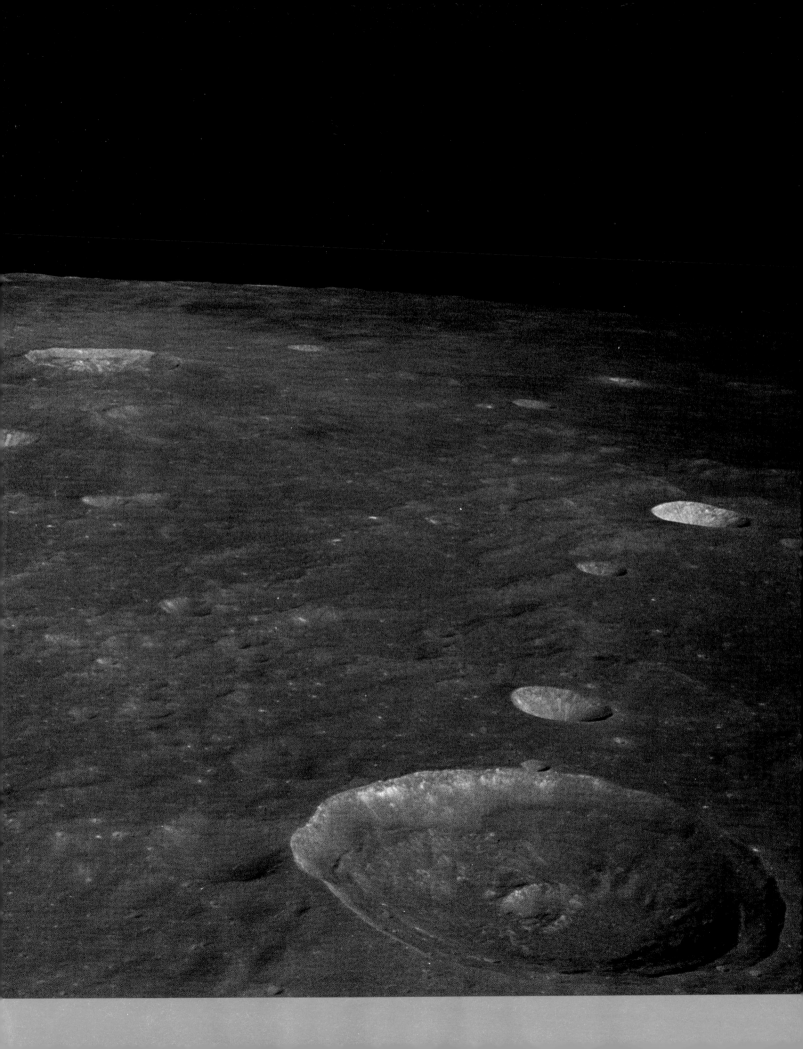

Alien on an alien world, Edwin "Buzz" Aldrin of *Apollo 11* emerges from his spindly craft to set up a foil detector for recording the solar wind.

Dwarfed by a split boulder, *Apollo 17* crewman Harrison Schmitt gazes into lunar hills stretching away from the Taurus-Littrow landing site.

23

**TIME LIFE** ®

*Other Publications:*
AMERICAN COUNTRY
THE THIRD REICH
THE TIME-LIFE GARDENER'S GUIDE
MYSTERIES OF THE UNKNOWN
TIME FRAME
FIX IT YOURSELF
FITNESS, HEALTH & NUTRITION
SUCCESSFUL PARENTING
HEALTHY HOME COOKING
UNDERSTANDING COMPUTERS
LIBRARY OF NATIONS
THE ENCHANTED WORLD
THE KODAK LIBRARY OF CREATIVE PHOTOGRAPHY
GREAT MEALS IN MINUTES
THE CIVIL WAR
PLANET EARTH
COLLECTOR'S LIBRARY OF THE CIVIL WAR
THE EPIC OF FLIGHT
THE GOOD COOK
WORLD WAR II
HOME REPAIR AND IMPROVEMENT
THE OLD WEST

This volume is one of a series that
examines the universe in all its aspects,
from its beginnings in the Big Bang to the
promise of space exploration.

# VOYAGE THROUGH THE UNIVERSE

# OUTBOUND

BY THE EDITORS OF TIME-LIFE BOOKS
ALEXANDRIA, VIRGINIA

# CONTENTS

# ARTH

With a gentle push, Alexei Leonov drifted away from his Voskhod space capsule. A hundred miles below him hung the Earth. An hour before, in the morning of March 18, 1965, Leonov and fellow cosmonaut Pavel Belyayev had blasted off from Baikonur Cosmodrome in the Soviet republic of Kazakhstan, climbing through lightly falling snow. Now, far above the cloud-whorled atmosphere, Leonov was making history as the first human to "walk" in space—a term that implied far more control than he actually had. Encased in a bulky pressurized suit, breathing oxygen from a life-support backpack, he sailed weightless, linked to his craft by a slender tether.

Leonov floated in the vacuum for a few minutes, drinking in the beauty of his home planet and the myriad stars beyond. Then he used the tether to pull himself back to the orbiting ship. The plan called for him to crawl back through the air lock to the capsule. Soviet engineers had assured him that this, like the rest of the excursion, would be easy. The engineers were wrong. Leonov found that his protective suit, stiffened by its internal air pressure, kept him from bending his waist sufficiently to get into the narrow air lock. Suddenly the dreamlike spacewalk turned to a nightmare: With only enough oxygen for an hour, he faced the grim possibility of slow suffocation just a few feet from safety. The alternative was to release some air from his balloonlike suit to make it more flexible—at the peril of a painful and possibly fatal case of the bends, a malady caused when reduced pressure allows dissolved nitrogen to bubble rapidly out of the blood. Battling panic, Leonov decided to take the risk. While flight controllers monitored his racing heart and ragged breath, the cosmonaut opened the valve that lowered the pressure in his suit to four pounds per square inch, barely a quarter of normal atmospheric pressure. Turning again to the air lock, he succeeded in maneuvering into the opening. As sweat rolled down his brow and fell burning into his eyes, he hauled the hatch shut behind him and quickly repressurized, thus avoiding the onset of the bends. The gamble had saved his life.

The misadventures of *Voskhod 2* had just begun, however. The next day, as the cosmonauts prepared to return to Earth, their autopilot mechanism failed, forcing Belyayev to assume manual control for the homeward leg. He had to take the craft through an extra orbit before the descent could begin, and although the passage through the atmosphere went off without a hitch, the landing was hundreds of miles away from the intended site. Instead of coming down on the open plains of Kazakhstan, near the Caspian Sea, the capsule landed far to the north in a snow-covered birch forest in the Ural Mountains.

Suffering from the cold and uncertain about when they would be rescued, Leonov and Belyayev clambered out of the craft. Still clad in their spacesuits, they kindled a fire and huddled beside the flames. With dusk came a scene from Russian folklore: Wolves prowled the woods, slowly circling toward the fire. The cosmonauts hastened back to their cramped cockpit, where they shivered through the night as the wolf pack clawed at the capsule's skin. Not until dawn did the wolves retreat and a rescue party arrive.

Centuries hence, when colonies sprout on the Moon and when Mars is a way station for journeys to Alpha Centauri and beyond, spacefarers may still recount the adventures of the *Voskhod 2* cosmonauts. Looking back on their species's first faltering steps toward the stars, they will be able to point out some archetypal themes of space exploration: beauty, danger, adventure, and above all, human courage and ingenuity.

## A VISION REALIZED

The dream of space travel was not new when Earth's inhabitants first broke through the planet's atmosphere in the 1960s. Those early, brief forays were the culmination of nearly a century of conceptual and technical achievement. Even as advances in astronomy made it clear that Earth occupies an insignificant corner of a vast cosmos, dedicated scientists and engineers pursued a vision of exploration that extended to the Moon, the planets, and—someday, perhaps—to remote solar systems that could only be guessed at.

As ventures into space became more confident, orbital missions, once measured in hours, stretched to days and weeks. Experience proved that humans could withstand the rigors of rocket launches and function effectively in weightlessness. Primitive capsules gave way to more sophisticated and spacious craft, their passengers now active pilots controlling ever more complicated missions. Just eight years after the pioneering orbital flights, American astronauts escaped Earth's gravity to walk on the surface of the Moon.

By the end of the 1980s, space explorers had mastered the fundamentals of survival in the hostile environment of space and were increasingly turning their attention to scientific studies. Biology, medicine, atmospheric science, and astronomy were among the disciplines that benefited from laboratories in space. Meanwhile, cosmonauts spent up to a year at a time on a Soviet space station in Earth orbit, learning lessons about human endurance that will be applied on the next great space adventure—a manned mission to Mars.

The workhorses of the dawning space age are the rockets that have lofted humans and hardware into orbit and beyond. In principle, a rocket is simple, the concrete expression of Newton's third law of motion, which asserts that every action engenders an equal and opposite reaction. When fuel burns in a rocket motor, it expands explosively. This expansion is directed rearward, producing force—called thrust—that propels the rocket forward. Unlike jet engines, which burn fuel mixed with air scooped from the atmosphere, rocket motors must carry with them all the ingredients needed for combustion. The propellants must include a fuel and an oxidizer; these may be in the form of liquids held in separate tanks or of premixed compounds of solid chemicals.

Although the underlying concept is elementary, building a rocket powerful and reliable enough to lift human beings into space is a prodigiously complicated business *(pages 25-29)*. Solid propellants must be precisely mixed and carefully packed; any imperfections could cause catastrophic explosions. Liquid propellants, which generate more power, require pumps, plumbing, and ignition apparatus that increase the likelihood of disastrous breakdowns. Steering a rocket to keep it aligned and on course calls for some way to divert the exhaust gases, or additional small steering

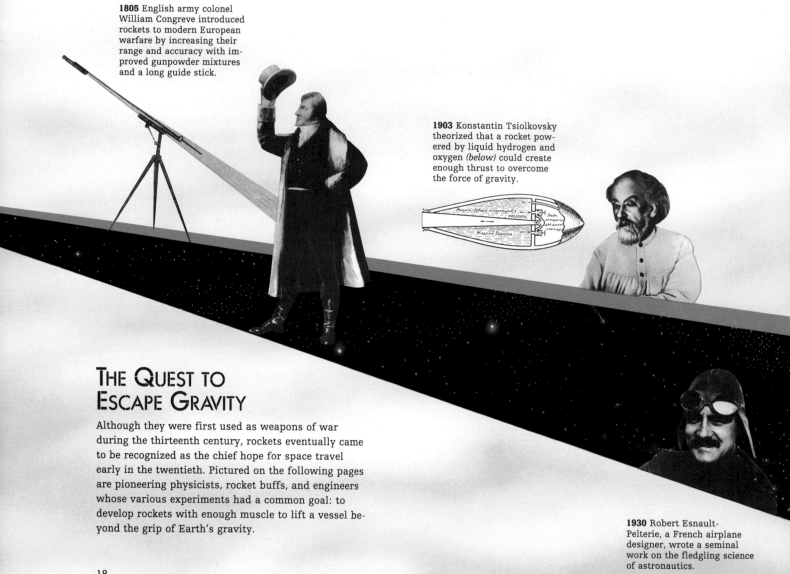

**1805** English army colonel William Congreve introduced rockets to modern European warfare by increasing their range and accuracy with improved gunpowder mixtures and a long guide stick.

**1903** Konstantin Tsiolkovsky theorized that a rocket powered by liquid hydrogen and oxygen *(below)* could create enough thrust to overcome the force of gravity.

# THE QUEST TO ESCAPE GRAVITY

Although they were first used as weapons of war during the thirteenth century, rockets eventually came to be recognized as the chief hope for space travel early in the twentieth. Pictured on the following pages are pioneering physicists, rocket buffs, and engineers whose various experiments had a common goal: to develop rockets with enough muscle to lift a vessel beyond the grip of Earth's gravity.

**1930** Robert Esnault-Pelterie, a French airplane designer, wrote a seminal work on the fledgling science of astronautics.

motors that can be fired periodically. These and other systems have been perfected only after countless trials, ending as often as not in flaming failure.

## AN ANCIENT TECHNOLOGY

The process of learning about rockets began many centuries ago. Rocketry can be traced at least as far back as 1232, when the defenders of the Chinese city of Kaifeng drove off attacking Mongol invaders with what they called "arrows of flying fire." These primitive devices, probably tubes packed with a gunpowder-like mix of charcoal, saltpeter, and sulfur, established the role of rockets for centuries: They were weapons, designed not just to rise above the Earth but to fall back with devastating effect. Because their trajectories were only vaguely predictable, they were usually fired in volleys to increase the chances of hitting a target. Rockets were adopted as siege weapons by Asian and European armies during the fifteenth century, but cannons offered more accuracy and gradually pushed them from the military stage.

Rocketry enjoyed a brief resurgence in the late eighteenth and early nineteenth centuries, then faded from the scene again as cannon performance advanced. By the end of the nineteenth century, the only serious thinking about rocket design came from visionaries less concerned with hitting terrestrial targets than with building spaceships that could leave Earth entirely. Inspired by novels such as Jules Verne's *From the Earth to the Moon* and H. G. Wells's *War of the Worlds,* young readers with a technical bent began to consider the feasibility of space flight. For many, the universe held the irresistible allure of an unknown sea awaiting exploration. It was human destiny, in the words of one, "to set foot on the soil of the asteroids, to lift

**1930** Hermann Oberth *(below, left)* and Klaus Riedel *(below, right),* speculating that rockets could launch payloads into orbit, successfully tested this liquid-fuel rocket engine.

**1932** Wernher von Braun was still a physics student in Germany when he tested this small oxygen-alcohol rocket. He later developed the world's first ballistic missile.

**1933** Friedrich Tsander *(left)* and Sergei Korolyov *(below, shown in the 1950s)* built the USSR's first liquid-propelled rocket, the GIRD-X *(right),* which reached an altitude of more than three miles.

by hand a rock from the Moon, to observe Mars from a distance of several tens of kilometers, to land on its satellite or even on its surface.''

Those words came from the pen of Russian writer and mathematician Konstantin Tsiolkovsky, acclaimed by many as the father of modern rocket science. Tsiolkovsky was born in 1857 in Izhevskoye, a farm village in central Russia, and as a child was rendered partially deaf by an illness. Denied a normal social life, he compensated by building a private world of futuristic technology. He taught himself mathematics and physics and proved such an apt pupil that he was able to leave the provinces to attend technical school in Moscow. Tsiolkovsky leavened his studies with the reading of novels, particularly, as he put it, those by ''that great fantastic author Jules Verne.'' Soon the schoolboy was dreaming of travels beyond Earth. One day, at the age of sixteen, he came up with the idea of a mechanism that would harness centrifugal force to power a spaceship. So excited by its implications that he could not sleep, he roamed the streets of Moscow all night. By morning he had discovered that his concept had a basic flaw, but the failure seemed only to sharpen his desire to uncover the secrets of space flight.

In the late 1870s, Tsiolkovsky took up a teaching post in Kaluga. During the next two decades, with virtually no public or private support, he carried out experiments in a makeshift laboratory and developed a theory of jet propulsion. At the same time, he began writing science-fiction tales with titles such as *On the Moon* and *Dreams of Earth and Heaven.*

In 1903, Tsiolkovsky published his findings in a paper filled with theoretical breakthroughs that would revolutionize rocketry. He derived a formula correlating a rocket's flight performance with the types and weights of its propellants, thus enabling later researchers to accurately predict the performance of any given rocket. In addition, he demonstrated mathematically that, with propellants superior to black powder, rockets could ascend beyond the atmosphere to orbit the Earth. He rightly understood that the key element in the design of high-performance rockets was increased thrust, produced by maximizing the velocity of the exhaust gases. This in turn required the use

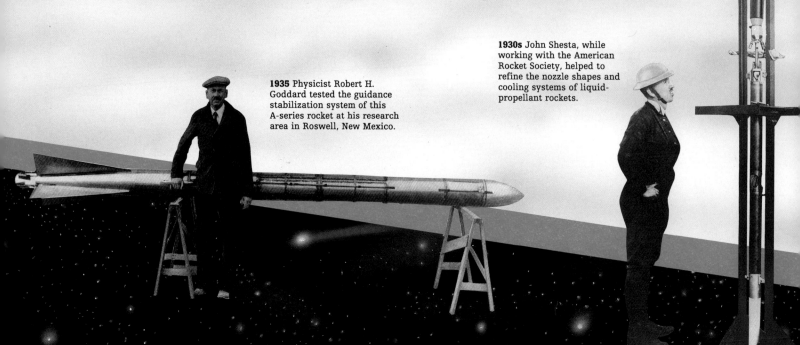

**1930s** John Shesta, while working with the American Rocket Society, helped to refine the nozzle shapes and cooling systems of liquid-propellant rockets.

**1935** Physicist Robert H. Goddard tested the guidance stabilization system of this A-series rocket at his research area in Roswell, New Mexico.

of potent fuels such as kerosene or liquid hydrogen, burned with liquid oxygen. Expensive and difficult to handle, these liquefied gases were just coming into industrial use at the time.

Few Russian scientists paid attention to Tsiolkovsky, who, as a low-ranking schoolteacher, had no standing in elite scientific circles. His fortunes changed after the 1917 Revolution, when the leaders of the new regime adopted the futuristic notion of space travel as part of their agenda. Tsiolkovsky, the most advanced theoretician in the field, became an eminent member of the newly chartered Russian Socialist Academy.

In his later years he examined ways to launch vehicles capable of making interplanetary voyages, concluding that only multistage rockets would deliver sufficient power. Subsequent investigations bore this out. In a modern rocket, the fuel and oxidizer together may make up all but about five percent of the total weight. Because it is wasteful to carry empty tanks after burning up their contents, the largest rockets are stacked in multiple sections, or stages, each with its own propellants and engines. Once the propellants in one stage are depleted, the section falls away and the engines of the next stage fire, delivering the additional thrust necessary to boost a payload into orbit.

## BUILDING ON THE VISION

Tsiolkovsky's visions of space rockets were still far from realization when he died in 1935. By that time, however, his theories were being tested by experimenters in several countries. One of the foremost of these architects of modern rocketry was an American named Robert H. Goddard. Born in Massachusetts in 1882, Goddard harnessed a passion for space travel—like Tsiolkovsky's, fed by the novels of Jules Verne—to an extraordinary mechanical handiness. As a youth, he often perched in a tree and imagined a rocket rising from the orchard below, bound for space. Years later, just such an orchard served Goddard as the launch site for the first liquid-propellant rocket, the progenitor of the massive boosters that would carry astronauts into space.

As a student at Worcester Polytechnic Institute in Massachusetts, Goddard

**1938** James Wyld, an American engineer, designed an engine that used fuel instead of water or air to cool itself.

**1947** Air Force Capt. Charles Yeager, piloting the Bell X-1 airplane, punched through the sound barrier in the first supersonic flight by a rocket-powered craft.

21

began experimenting with solid propellants, earning a reprimand from school authorities when one of his gunpowder rockets accidentally exploded in the basement of the physics building. After graduating in 1910, Goddard contracted tuberculosis, and during his long convalescence, he gave free rein to his infatuation with rockets, developing ideas that led to the issuance of two patents in 1914. One was for a multistage rocket design; the other was for rockets using both liquid and solid propellants. They were the first of eighty-three patents that Goddard would receive during his lifetime. After his death, his notebooks were mined for 131 more.

In 1914, hired as a physics instructor by Clark University, also in Worcester, Goddard turned his attention to more powerful and efficient liquid-propellant rockets, building a series of ever more complex engines. In 1916, he fired off a stationary motor inside a vacuum chamber, thus proving that rockets would function in space. Shortly thereafter, he wrote a paper for the Smithsonian Institution entitled "A Method of Reaching Extreme Altitudes." The altitudes Goddard was considering were indeed extreme: He proposed a program that would take a test rocket to the Moon. Neither Goddard nor the Smithsonian harbored illusions about the difficulty of such a project, but the institution granted him $5,000 to begin developing rocket hardware.

By the early spring of 1926, Goddard was ready to try to launch a liquid-propellant rocket. It was a rickety affair, cobbled together from pipe and scrap metal. The combustion chamber and exhaust nozzle were mounted at the top of the assembly, ahead of the propellant tanks. The fragile tubes that connected the two parts of the rocket doubled as supply lines for the gasoline and liquid oxygen that would be burned to generate thrust.

On March 16, amid the bare trees of an orchard near Worcester, Goddard ignited the engine from an alcohol stove placed below it. As his wife Esther recorded the momentous event with a movie camera, a flame blazed out of the exhaust nozzle. At first the rocket merely trembled on its stand; the camera, loaded with enough film for seven seconds, recorded only stasis. Then, twenty seconds after ignition, the engine's thrust finally exceeded the diminishing weight of the tanks and propellants, and the rocket struggled into the air. "It looked almost magical as it rose," Goddard commented at the time, "as if it said, 'I think I've been here long enough, I think I'll get the hell out of here.'" The rocket rose 41 feet and came down 184 feet from the launch stand.

Over the next three years Goddard continued to develop and launch liquid-propellant rockets. His fourth, in July 1929, was the first to carry instruments—an aneroid barometer, a thermometer, and a camera, all of which were recovered after the rocket plummeted back to Earth. This successful test had unwelcome results: Some uninformed observers mistook the noise and smoke of the rocket for an airplane crash, and the farm was quickly overrun with ambulances, police, and reporters. The state fire marshal imposed a ban on further flights, and Goddard's research prospects looked dim.

Eventually, however, the publicity attracted the attention of Charles Lindbergh, world famous for his solo transatlantic flight in 1927. Through Lindbergh's influence, Goddard began to receive grants from philanthropist Daniel Guggenheim, who contributed heavily to early aeronautics research. The funds allowed Goddard to take leave from his position at Clark University and turn his full attention to rockets and high-altitude research. From 1929 to 1941, Guggenheim and his foundation underwrote more than $150,000 in expenses. Goddard moved his base of operations to an isolated ranch not far from the town of Roswell in New Mexico's Eden Valley.

From the desert of Mescalero Ranch rose an assortment of rockets, built by Goddard and a select team of helpers. In 1935, one of the rockets exceeded the speed of sound—an aeronautical first. Another climbed to an altitude of 9,000 feet, higher than any known rocket flight. Other rockets exploded before leaving the launching tower, and a seemingly infinite array of malfunctions caused many postponements. Through it all, Goddard retained his equanimity; after one disastrous test, he revealed his disappointment only by muttering, "Well, there goes $10,000 up in smoke."

Trial and error led to steady improvements in design. Goddard built pumps powered by compressed nitrogen to supply fuel to the rocket engines. Gyroscopic guidance systems steered the missiles, either by turning vanes in the exhaust or by turning the engine itself. The engines were cooled by circulating the fuel and oxidizer around them. (Oxygen, for example, is liquid only below minus 297 degrees Fahrenheit.) In fact, virtually all the essential elements of modern liquid-propellant rockets first appeared in the missiles Goddard launched from the New Mexico desert.

## A MORE PUBLIC APPROACH

Goddard's work in near seclusion contrasted starkly with early rocket research in Germany, where a contemporary of Goddard's relied on publicity to fund his efforts. While serving in the German Army during World War I, Hermann Oberth mapped out plans for a high-altitude rocket, carefully engineering every detail of the hypothetical missile. After the war he expanded his analysis for a doctoral dissertation in astronomy at the University of Heidelberg. The conservative Heidelberg faculty rejected Oberth's effort because it touched on a variety of subjects rather than sticking to the narrow line of astronomy. Undaunted, Oberth revised the manuscript and in 1923 published it as *The Rocket into Interplanetary Space.*

The book contained hard figures on rocket dynamics and came with a foldout rendering of a two-stage liquid-propellant missile complete with gyroscope guidance system. The preface, which amounted to a rocketeering manifesto, sparked controversy among scientists by claiming not only that space travel was possible but also that it would someday be commercially profitable. Oberth's exhaustive study of rocketry was reprinted for years, becoming the bible for aerospace engineers.

Without capital, Oberth could do nothing to test the merit of his ideas, so he moved to Rumania to take up a career as a schoolteacher. Then, in 1928, an odd opportunity presented itself: The innovative German filmmaker Fritz

Lang hired Oberth as a scientific adviser for *Girl in the Moon,* a movie set in outer space. Although *Girl in the Moon* took many liberties with the facts as they were then known, its treatment of rocketry was largely accurate. In a scene concocted for the movie, Oberth's scale-model rocket—correct to the last detail—blasted off at the end of a dramatic countdown. The make-believe ship followed a figure-eight flight path to the Moon and back, a course that would be traced in reality four decades later by the *Apollo 8* mission.

As a publicity stunt, Lang's studio commissioned Oberth to build a functioning rocket, to be launched on the night of the film's premiere. Oberth was far too optimistic when he accepted the assignment; when the deadline arrived, he had not even been able to test the engine. The unfinished rocket was relegated to a dusty corner of a workshop, but he continued working on rockets with a small stipend from the German government. He was joined by enthusiastic members of a recently formed club, the Verein für Raumschiffahrt (VfR), or Society for Space Travel.

Oberth's bright young assistants provided the nucleus of a team that conducted rocket research in Germany through the early 1930s, even after Oberth returned to his teaching post in Rumania. Their ranks were swelled by unpaid technicians unable to find work in depression-racked Germany. Headed by Rudolf Nebel, an engineer who also proved to be a master scavenger, the team carried out a series of test flights at an abandoned industrial compound on the outskirts of Berlin. Their rocket, like Goddard's earliest designs, had a front-mounted engine that ran on gasoline and liquid oxygen. The slender apparatus eventually attained a height of 3,300 feet. Unfortunately, the parachute that was supposed to float the spent missile back to the surface failed to open, and the rocket was smashed.

The VfR engineering team spent four hardscrabble years seeking performance improvements. They devised a way to dissipate the extreme heat generated by the burning fuel, by wrapping the combustion chamber in a jacket of tubes filled with cooling water. Their engines burned relatively steadily and produced substantial thrust for their size. But the VfR failed to solve a number of other technical problems, with the result that the rockets steered poorly, flew unpredictably, and never got very far off the ground.

**ENTER THE MILITARY**

In 1934, the VfR disbanded, its meager budget having proved unequal to the rising cost of experimental hardware. By that time, however, the German Army had begun to show an interest in rockets, perceiving them as a way to outflank restrictions on long-range artillery imposed by the Treaty of Versailles at the end of World War I. The army's effort was organized by Walter Dornberger, a young captain in the Weapons Department, who took in several veterans of the VfR research program, including Wernher von Braun, a star apprentice of Oberth and Nebel. Von Braun conducted his first investigations for the army in 1932 at the deserted Kummersdorf Proving Ground, an artillery range about forty miles south of Berlin. There, in a makeshift labo-

**5** At about sixty miles up, the main engines are cut off; the external tank is jettisoned, disintegrating as it falls.

# RIDING ROCKETS INTO SPACE

**4** The shuttle continues to ascend, its trajectory now guided by adjusting the position of the main engine nozzles.

Developed chiefly for military purposes during the first half of the twentieth century, rockets gave birth to the space age when they began boosting human beings into orbit in 1961. In the three decades since then, rocket technology has undergone further refinement, but the underlying principle of rocket propulsion remains the same: Fuel of some sort is burned in a chamber fitted with a nozzle at one end; hot gases escape through the nozzle, creating an opposite reaction—forward momentum—at the other end of the chamber *(box, below)*.

As described here and on the following four pages, the rockets used by the U.S. space shuttle burn both liquid and solid fuels during launch and liquid fuels for maneuvers in orbit. Liquid fuels can be explosive if mixed improperly or stored at too high a temperature, but they allow a rocket engine to be throttled up and down, like the engine of a car. Solid fuel is less volatile than liquid fuels and easier to store and transport. But solid rocket motors are less useful for orbital maneuvers because they cannot be throttled; once ignited, solid fuel burns until it is gone.

**3** The spent SRBs are jettisoned after two minutes and parachuted to Earth, where they are retrieved for reuse.

**2** As the shuttle rises, atmospheric pressure decreases and exhaust plumes expand.

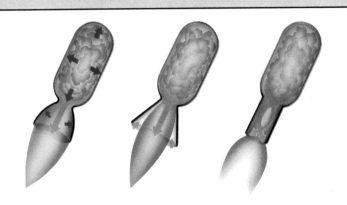

Hot gases produced in a rocket's combustion chamber push equally on all its sides *(red arrows)*. Allowing the gas to escape *(pink arrow)* transforms the opposing pressure into forward movement, or thrust. The most efficient nozzle *(left)* is contoured to the exhaust stream, allowing the escaping gas to expand just enough to fill the nozzle. A nozzle that lets the gas expand too much *(middle)*, or one that prevents it from expanding *(right)*, wastes the energy and thrust potential of the exhaust stream.

**1** Two SRBs (solid rocket boosters) and three liquid main engines generate a full 7.7 million pounds of thrust.

# TWO MINUTES OF SOLID THRUST

In the two minutes before they are jettisoned, the shuttle's two solid rocket motors burn propellant at the rate of about ten tons per second, generating 6.6 million pounds of thrust to overcome Earth's gravity. This enormous power is produced by the interaction of two key ingredients—a fuel and an oxidizer, which allows the fuel to combust—held together by an inert binding agent. For the shuttle's solid propellant, aluminum is the fuel, ammonium perchlorate is the oxidizer, and a polymer binds the two.

Loading propellant is a relatively straightforward task. The components are stirred up like a batter and poured into rocket casings that each have a removable core of variable design. After the mixture cures into a firm, rubberlike mass called a grain, the core is removed, leaving one or more perforations in the grain. When the propellant is ignited, only the grain's exposed surfaces burn. Tailoring the shape of the grain and the total area and contours of its exposed surfaces gives it a special pattern of burning that affects the rate of gas production and the generation of thrust.

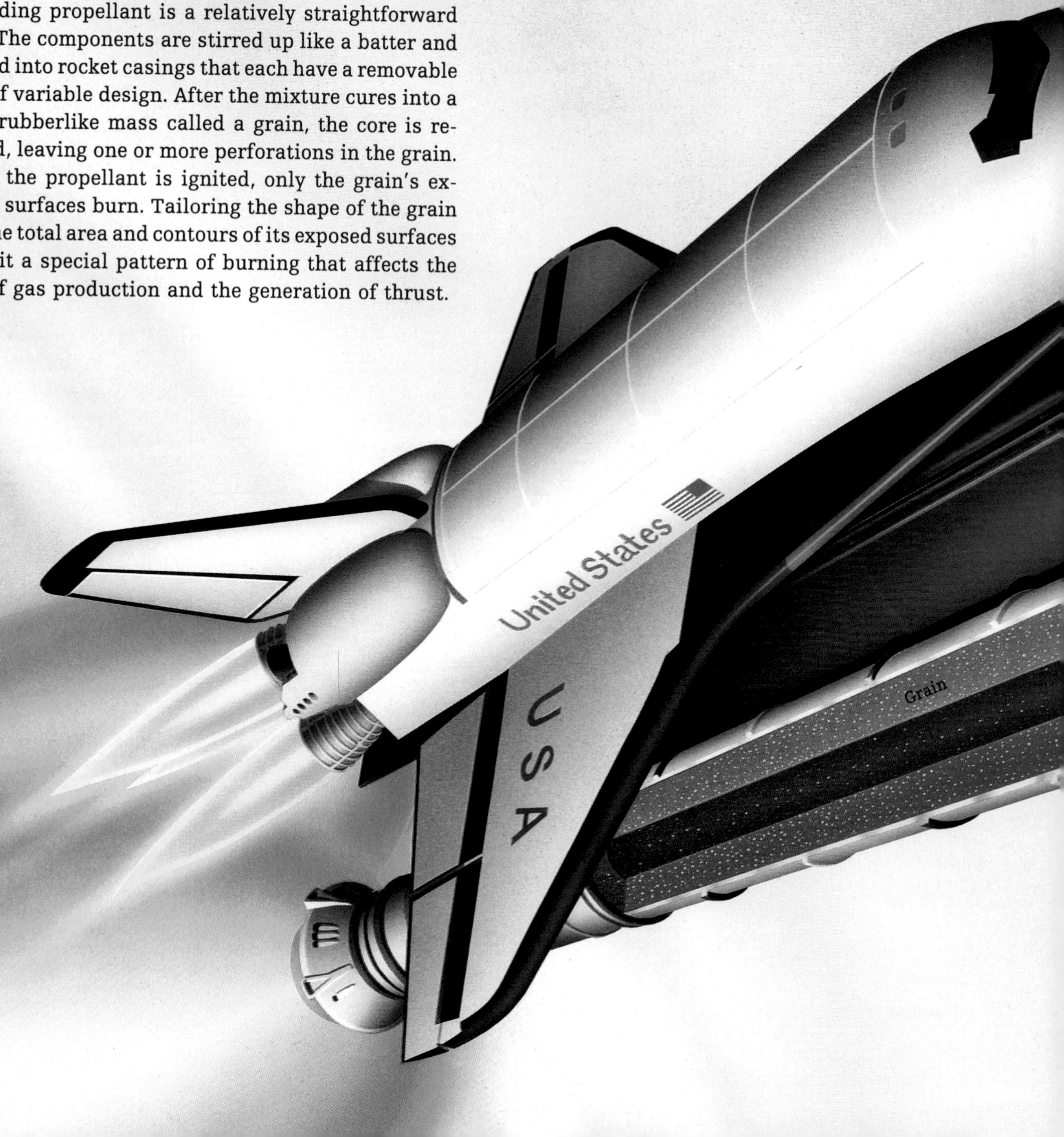

United States

USA

Grain

Separation Motors

11 Point Star

Parachutes

Igniter

40 Point Star

30 Point Star

The solid grain's geometry determines the area and contours of its exposed surfaces, and thus its burn pattern. From left to right above: Consumed in about twenty seconds, an eleven-point star used in the shuttle contributes to maximum thrust at the beginning of the burn. The so-called end burner produces constant thrust throughout the burn. A cruciform grain yields progressively less thrust, and the multiport grain produces high thrust for a short time.

Located at the top of the grain is an ignition system made up of two separate grain configurations. An electric current triggers a chain reaction that ignites a thirty-point star, then a forty-point star, which in turn almost instantaneously lights the entire exposed surface of the main supply of solid fuel.

27

In the first stage of combustion, liquid hydrogen *(blue)* and liquid oxygen *(yellow)* are pumped through a low-pressure *(1)* and then through a high-pressure *(2)* turbopump before being injected into two preburners *(3)*, where they are partially burned. The resulting fuel-rich, gaseous mixture *(green)* then enters the combustion chamber *(4)*, along with additional liquid oxygen *(5)*, to set off the second stage of combustion. The two-stage system allows the three shuttle engines together to burn propellant at the rate of a thousand gallons per second.

Orbital Maneuvering and Reaction Control System

United States

Main Engines

Liquid Oxygen

Liquid Hydrogen

# A LIQUID-FUEL WORKHORSE

Once a craft is off the launch pad, liquid rocket engines offer two primary advantages over solid rocket motors: They extract a higher ratio of thrust per pound of fuel, and that thrust can be regulated during flight. Simply described, the shuttle engine system employs a highly volatile liquid fuel and an oxidizer, which are combined under high pressure in a combustion chamber and ignited. Some of the energy released keeps combustion going as more fuel and oxidizer are injected into the chamber, producing high-pressure gases that give the rocket its thrust. By controlling the propellant flow, thrust may be increased or decreased to prevent a craft from accelerating too fast and subjecting both craft and crew to dangerous forces. The shuttle's external tank *(above)* can carry just under 400,000 gallons of liquid hydrogen and about 140,000 gallons of liquid oxygen, which are combined in a two-stage combustion process *(diagram, opposite, top)* in each of the craft's three main engines.

Although improvements in engine design and more powerful fuel combinations are expected to make the liquid rocket engine still more efficient, liquid fuels may eventually be supplanted. Possible successors include such high-energy power sources as nuclear reactions, solar energy, and ion propulsion.

ratory that was little more than a concrete pit with a sliding roof, von Braun and his assistant, mechanic Heinrich Grunow, drove themselves hard, striving to build a sample missile in record time.

The technology of the day was sadly lacking: To have any chance of hitting a remote target, a rocket would need, among other things, better gyroscopic guidance, automated controls for cutting off its motor, and more efficient fuel pumps. Von Braun's first concern, though, was to build a satisfactory engine, a task he accomplished in short order. By January 1933, he had supervised tests on the Aggregate-1 rocket, or A-1. Nearly a year later, the A-2, fitted with a new engine, took off from a test pad on Borkum Island in the North Sea and flew to 7,900 feet, briefly capturing the high-altitude record.

In 1936, the army stepped up its rocket program, installing von Braun and an expanded team of scientists and engineers, along with their families, in the fishing village of Peenemünde on the island of Usedom in the Baltic Sea. Terse and occasionally argumentative, von Braun nonetheless won the confidence of his subordinates with fairness and group spirit. Within a few years, he was supervising a site packed with specialized laboratories, liquid-propellant plants, supersonic wind tunnels, engine testing stands strong enough to endure thousands of pounds of thrust, and concrete launching pads.

Von Braun unveiled his next creation, the A-3, in 1937. With a powerful alcohol and liquid-oxygen engine, it stood twenty-two feet tall and had a gyroscopic guidance-control system for adjusting direction in three planes. But the A-3 was a miserable bust. On its first test flight, it flared and rose from the pad for five seconds; then its parachute opened into the exhaust jet and the rocket tumbled into the Baltic. The second test rocket was fired without a parachute. When it also crashed, the engineers concluded that the complicated gyro system was at fault and began a redesign effort.

At this point, with no useful weapon in sight, the army lost patience with the program, which had produced only expensive test missiles. Dornberger received a command from headquarters: Henceforth, the team must focus strictly on building a field weapon that would carry a large warhead much farther than was possible with long-range artillery. Von Braun immediately looked into the feasibility of scaling up the next missile in the series. It was already on the drawing board but, like its predecessor, had a planned range of scarcely ten miles and could boost a payload of about 150 pounds—enough for a metal canister to hold the missile's recovery parachute.

The rocket underwent a complete overhaul. Scientists working with the Peenemünde wind tunnel studied the tail fins and reshaped them to withstand supersonic conditions. Guidance and control technicians contrived a method for steering with heat-resistant graphite vanes that deflected the exhaust gases. Engineers perfected steam-turbine pumps to feed alcohol and liquid oxygen into the engine at high speeds. Finally, a massive engine was forged, with an enlarged combustion chamber in which fuel would burn with 95 percent efficiency. The engine was so gargantuan, with a thrust topping 55,000 pounds, that the engineers were afraid even reinforced stands could not harness it for static trials. Instead, they tested several smaller units and made their preliminary evaluations of the new rocket by proxy.

The finished rocket, designated the A-4, was forty-seven feet tall, with a takeoff weight of about 28,000 pounds; one ton was reserved for a high-explosive warhead. On the first test, in the spring of 1942, a fuel line clogged and the rocket plunged into the Baltic barely half a mile from the pad. A second rocket fared little better: Uncontrollable vibrations caused it to splinter in midair less than a minute after takeoff and fall into the sea five miles away.

## THE FIRST SUCCESS

All summer the team scrambled to correct flaws, and on October 3, 1942, a third specimen stood poised on the launch apron. Decorating its tail was a cartoon of a girl, a rocket, and a crescent moon painted by experimenters who had once thrilled to the fantasies of Fritz Lang. Among the spectators was Hermann Oberth, now working on advanced projects at Peenemünde. Dornberger later recalled the launch: "It was an unforgettable sight. In the full glare of the sunlight the rocket rose higher and higher. The flame darting from the stern was almost as long as the rocket itself." Accelerating to a speed of 4,400 feet per second, it quickly passed out of sight. Primitive radio telemetry allowed controllers to monitor such vital signs as the timing of valve openings and the temperature of the engine wall. All went well: The rocket rose to an altitude of 52 miles before plunging back to Earth 116 miles away.

Dornberger's elation was boundless. "Today," he proclaimed, as he shook hands with Oberth, "the spaceship has been invented." At a victory party that evening, members of the A-4 team—many of them veterans of the VfR—celebrated the dawn of a new era. But if its creators saw the rocket as a spaceship, the army continued to regard it as a weapon of war. Dubbed the V-2 (for Vengeance Weapon Number 2), it was intended to devastate Germany's enemies. Lethal, if somewhat indiscriminate—the missile was accurate to within about fifteen miles—the V-2 was useful only against large cities. Beginning in September 1944, volleys of V-2s were launched toward England. Most struck in and around London, killing and wounding thousands of people.

As the war drew to a close, Dornberger and von Braun began to speak publicly about the peacetime possibilities of the rocket as a vehicle for space exploration. Von Braun's remarks were tinged with regret that the development of rocket-powered flight owed so much to the gods of war. Such sentiments earned him the enmity of Heinrich Himmler, chief of the SS, Hitler's central security organization. Himmler had the scientist arrested on trumped-up treason charges, which were dropped only after Dornberger, by now a major general, intervened.

Allied intelligence kept a close watch on the German V-2 effort, and when the conflict in Europe ended in the spring of 1945, the victors hastened to grab the spoils. American troops seized the Mittelwerk, a V-2 fabrication plant, taking possession of parts for nearly a hundred rockets. The wind tunnel from Peenemünde was shipped to White Oak, Maryland, in pieces. Thirteen years' worth of records, cached by von Braun in a mine shaft, also went to the United States. The Soviets, meanwhile, discovered two fully assembled V-2s in Poland, complete with service manuals. These were shipped to the USSR.

Convinced that he would fare better with the Americans than with the Red Army, von Braun made his way south by train, accompanied by more than 500 members of his staff and their families. Eluding both Soviets and Germans for more than two months, the group finally surrendered to U.S. forces in Bavaria. Thus a large part of the German rocket establishment reached the United States intact, resuming activity in September 1945 at Fort Bliss, Texas. Soon thereafter, they began testing V-2s at White Sands, New Mexico—not far from Robert Goddard's test site at Mescalero Ranch.

Goddard, who died in August 1945, never saw the new missiles roaring into the desert skies. His rockets had anticipated the V-2s, but the Germans had indisputably advanced the technology further in a decade than it had come in the previous 700 years. The V-2 was a relatively reliable vehicle, capable of hurling a substantial payload above the atmosphere for a brief journey at the edge of space. Now von Braun and his coworkers hoped to use it as the basis for true space rockets, powerful enough to place satellites in orbit. The final stage of a satellite booster would have to achieve a velocity of 18,000 miles per hour—nearly six times that of the V-2.

By 1949, the White Sands research group had put together a three-stage rocket that set a record for penetration into space. The vehicle was assembled by mounting a rocket known as the WAC-Corporal atop a V-2. The WAC-Corporal burned a unique propellant, a combination of liquid chemicals that spontaneously combusted when mixed, eliminating the need for an ignition device. When that fuel was exhausted, a solid-fuel top stage ignited, speeding upward to an altitude of 250 miles. Von Braun and his colleagues longed to take the next step and launch a satellite. But space was not a military priority. They would have to wait another ten years.

## PROGRESS IN RUSSIA

The Soviet Union, meanwhile, had put 6,000 captured German technicians to work alongside Soviet engineers soon after the end of the war. The engineers had been secretly developing rockets under state auspices for more than a decade, following in the wake of work by Friedrich Tsander, a Lithuanian who during the early part of the century had corresponded with Tsiolkovsky. After Tsander's death in 1933, the leadership in rocket research passed to Sergei Korolyov, a test pilot turned designer, and within a few years, Soviet rockets had reached altitudes of six miles, nearly four times as high as Goddard's best shot. In the late 1930s, however, progress halted as Korolyov, along with hundreds of other engineers, was banished to the Siberian gulag in the course of Joseph Stalin's political purges. The rocket engineers were finally released in 1946, and in the postwar years, Korolyov oversaw the production of thousands of V-2s. The Soviet Union, plunging into a furious arms race with its erstwhile ally the United States, also began to develop a much larger missile,

Trailing antennas like four spidery legs, Soviet satellite *Sputnik 1* appeared in the skies on October 4, 1957. The diminutive object, no larger than a beach ball, orbited the Earth for three months.

the Pobyeda (Victory), which could throw a heavy warhead over a range of 900 miles. In 1949, the Pobyeda went into secret production.

In the United States, funding for rocket development was much harder to come by. Von Braun and his team, transferred to the Redstone Arsenal in rural Huntsville, Alabama, were instructed by the army to build the Redstone missile, a kind of dressed-up V-2 capable of carrying a nuclear warhead. Steady budget cuts in the early 1950s slowed the project, which had little support in Congress. The first flight, in August 1953, ended with a crash only four miles from the launch pad, but soon Redstones were rising regularly from the missile test facility at Cape Canaveral, Florida.

Official interest in space was rekindled in 1954, when the Office of Naval Research began to look into the idea of launching a satellite to assess the military use of near-Earth space. Support for the launch of artificial satellites had been strengthening among scientific organizations ever since a group of influential scientists in 1950 had proposed a year-long internationally coordinated study of the upper atmosphere. Capitalizing on the interest surrounding this so-called International Geophysical Year, von Braun and others lobbied for a satellite project that would use existing technology and thus cost relatively little. In July 1955, President Eisenhower yielded to the pleas and announced that the U.S. would put a small satellite into orbit in time for the IGY, which was to run from July 1, 1957, to the end of 1958. A jurisdictional dispute between the army and the navy ensued: Each proposed its own combination of launcher and satellite. The army championed von Braun's Project Orbiter, which was essentially a Redstone topped by three solid-propellant stages, while the navy campaigned for Vanguard, a new three-stage vehicle. In August, an advisory committee voted for the Vanguard.

In Huntsville, the news was a bitter pill. The month before, a gray-haired man with beret and briefcase had arrived in Huntsville to work with the Redstone team: Hermann Oberth, the mentor of many of the German scientists. Now it seemed that Oberth and his disciples would have no part in realizing the dream that many had pursued for nearly thirty years. Von Braun, however, was able to store several modified Redstones, now called Jupiter Cs, at Cape Canaveral. On September 20, 1956, the first of these rockets soared to an altitude of 682 miles before splashing down in the Atlantic 3,355 miles south of Cape Canaveral. It carried a dummy fourth stage filled with sand, where a satellite might have been. On that morning, von Braun later recalled, "we knew that with a little bit of luck we could put a satellite into space. Unfortunately, no one asked us to do it."

By this time, the Soviet Union was edging into the race to orbit. For several years, scientists and engineers had discussed the possibility of a satellite launch. Eager to demonstrate his nation's technological prowess, Nikita

Khrushchev announced in the summer of 1957 that the USSR, too, would put up a satellite for the International Geophysical Year. This was not mere braggadocio: At the new Baikonur Cosmodrome, Korolyov's team was readying a true intercontinental ballistic missile. The R-7 was a so-called parallel-stage rocket, with twenty engines divided among a central cylinder and four peripheral pods; the pods would peel off when their fuel was exhausted. Designed to carry a two-ton warhead over a range of 4,000 miles, the R-7 first flew in August 1957. After a few more successful launches, the missile was designated to carry the first Soviet satellite into space.

Korolyov hoped to time the launch to coincide with the hundredth anniversary of Tsiolkovsky's birth on September 17. With only a few weeks to prepare both the launcher and the satellite, he pushed his crew to the limit. Taking up residence in a small frame house a half-mile from the launch pad, he seldom slept, and his workers knew that he might drop in on them at any hour of the day or night. As technicians assembled the sections of the rocket, machinists toiled nearby to put together a simple satellite, an aluminum alloy sphere about twenty-three inches in diameter and fitted with two radio transmitters and four antennas. Despite virtually nonstop work, however, the deadline proved impossible to meet.

## THE DAWN OF A NEW AGE

Finally, on October 4, the rocket—affectionately dubbed "Old Number Seven"—stood ready on the launch pad. Technical problems had delayed the launch until well after sundown. When the engines ignited, observers were dazzled by the glare. Then the rocket lifted off, rising on an arching trajectory to the northeast, becoming a pinpoint of light in the sky. Two hundred miles above the surface, the central stage reached a velocity of 18,000 miles per hour and released its 184-pound payload, dubbed Sputnik (Fellow Traveler)—the first object built by humans to orbit the Earth. Like some small creature that had broken free of a confining shell, it came to life, its radios broadcasting a steady "beep beep beep" that reported its onboard temperature.

In the United States, politicians were shocked and rocket engineers chagrined. "We knew they were going to do it," exclaimed von Braun, momentarily losing his aplomb. "For God's sake, turn us loose and let us do something. We can put up a satellite in sixty days." In actuality, it took the now-frantic Americans eighty days. A first attempt on December 6, using the navy's Vanguard, failed when the rocket blew up on the pad. Humiliated worldwide—"Oh, what a Flopnik!" cried the London *Daily Herald*—the U.S. turned to the army's Jupiter C. It was quickly outfitted with a satellite named *Explorer 1*, a slender tube nearly seven feet long with a nose cone that held two radios, two antennas, and eleven pounds of instruments for measuring cosmic rays and other phenomena. On January 31, 1958, von Braun's space rocket finally flew. It was a perfect launch. Less than seven minutes after liftoff, *Explorer 1* was in orbit. The race—with a starting line on Earth and a finish line somewhere in the vast reaches of the universe—had truly begun.

# THE GRAVITATIONAL RULES OF THE GAME

During the late nineteenth and early twentieth centuries, rocket engineers like Russia's Konstantin Tsiolkovsky and Robert Goddard of the United States dreamed of building devices that could perform a thrilling role—lofting a craft beyond the bonds of Earth's gravity and into the reaches of space. Because the laws of planetary motion and gravitation had been worked out much earlier, Tsiolkovsky and Goddard fully understood the scope of the challenge they faced. The gravitational attraction between objects is universal; the more massive the object, the stronger its pull. In the case of rockets trying to blast free of Earth, the planet was for a long time the hands-down winner. As late as the 1930s, propulsion systems were too weak to achieve altitudes of more than a few miles.

Not until the late 1950s did improvements in liquid-fuel and solid-fuel rockets *(pages 26-29)* allow space engineers to begin launching vessels into Earth orbit for missions of varying duration. Soon they used Earth orbit, where the gravitational pull of the planet and the kinetic energy of the craft were perfectly balanced, as a springboard to the Moon, and the comings and goings imagined by the pioneers of rocketry became almost routine.

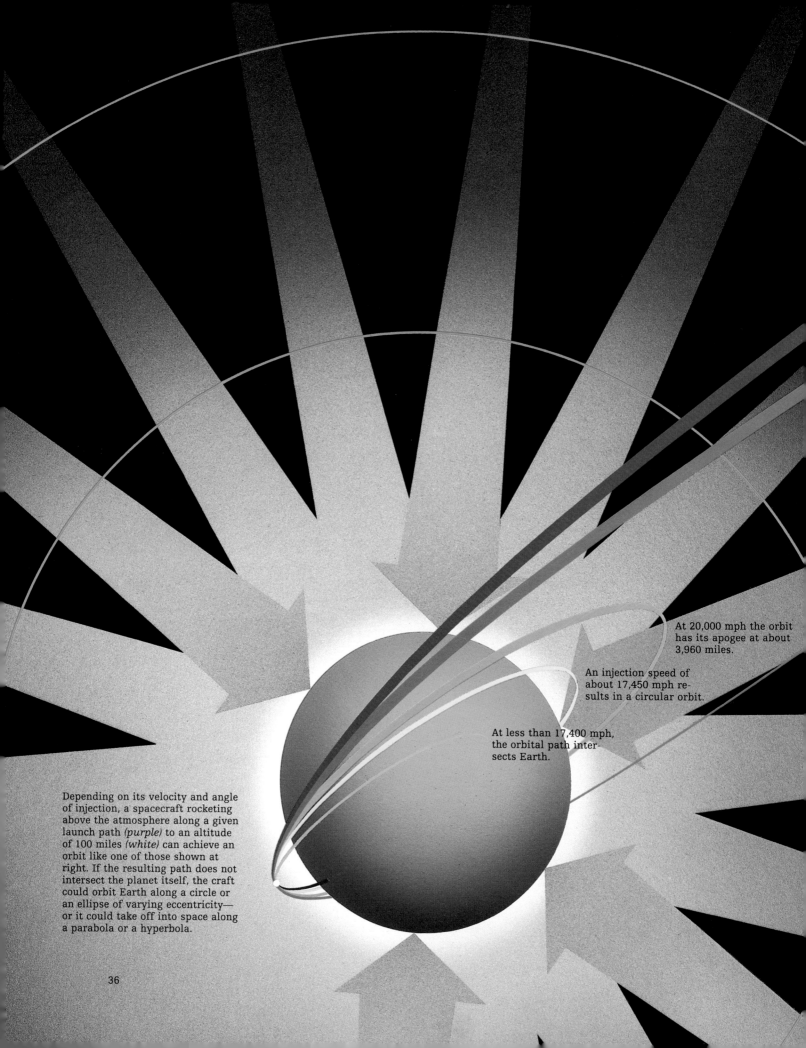

At 20,000 mph the orbit has its apogee at about 3,960 miles.

An injection speed of about 17,450 mph results in a circular orbit.

At less than 17,400 mph, the orbital path intersects Earth.

Depending on its velocity and angle of injection, a spacecraft rocketing above the atmosphere along a given launch path *(purple)* to an altitude of 100 miles *(white)* can achieve an orbit like one of those shown at right. If the resulting path does not intersect the planet itself, the craft could orbit Earth along a circle or an ellipse of varying eccentricity— or it could take off into space along a parabola or a hyperbola.

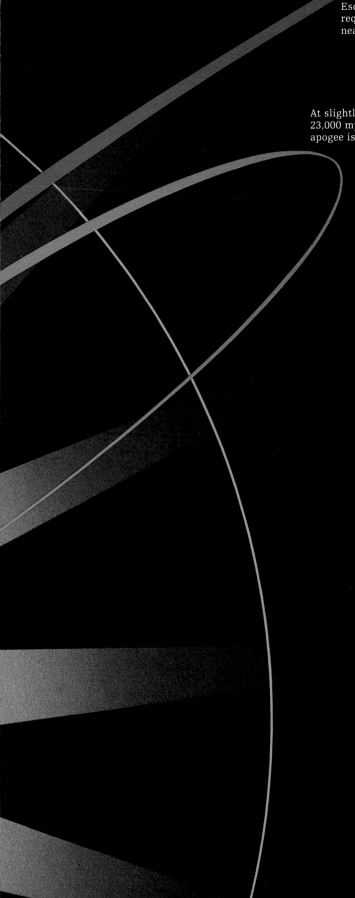

Escaping Earth's pull
requires injection at
nearly 25,000 mph.

At slightly less than
23,000 mph, the orbit's
apogee is 22,300 miles.

Earth's gravitational pull extends in all
directions *(gray arrows)*, but it weak-
ens with distance from the center of the
planet according to the so-called
inverse-square law. A spacecraft 7,920
miles above the surface *(inner blue circle)*, three times the distance from the
center to the surface, feels one-ninth the
surface gravitational attraction.

# OF SPEED AND ORBITS

A spacecraft orbiting the Earth (or any other celestial
body) is engaged in a delicate balancing act: To
achieve and then maintain its orbit, it must move fast
enough to counter the planet's gravitational attraction, which seeks to pull the craft to its center. This
attraction decreases with distance from the center of
the planet according to a relationship called the
inverse-square law. For example, a spacecraft orbiting about 15,800 miles above the surface *(outer blue
circle)* is five times as far from the center of the planet
as a craft ready to launch from the surface, 3,960 miles
from the center, and feels one twenty-fifth the gravitational pull of its surface counterpart.

   To go into orbit, a craft needs enough energy first to
rise above the atmosphere to the desired altitude and
then to boost itself at just the right speed and angle to
achieve an orbit of the desired shape. A craft injected
into orbit at an altitude of about 100 miles (exaggerated at left for clarity) could follow a number of orbital
paths, some leading far out into space. Once in orbit,
a craft's speed is related to its altitude; the farther it
is from the accelerating effects of Earth's gravity, the
slower its orbital velocity.

# AS THE WORLD TURNS

A spacecraft on the launch pad is in motion even before it lifts off, moving from west to east as the Earth rotates. For the most part, mission controllers use this running start to conserve fuel by launching the craft along a west-to-east path. Launching north or south into an orbital plane perpendicular to the direction of Earth's movement would reap no rotational benefits, and launching from east to west, against the planet's motion, would cost extra in energy. The ideal launch would take place from the equator, where surface speed is greatest, into an equatorial orbit. Since existing launch sites are many degrees of latitude above the equator, however, space engineers must do the best they can, launching their craft into an orbital plane that is inclined to the plane of the equator at an angle equal to the latitude of the launch site *(right)*.

Because perfectly circular orbits are difficult to attain, requiring a precise velocity and angle at injection, most orbits are elliptical, and the craft is always speeding up or slowing down *(below)*.

Spacecraft launched from Florida's Cape Canaveral travel from west to east to get a boost from the Earth's 935 mph surface velocity at that latitude. The plane of their orbit is inclined about 28.5 degrees to the equator. The point where the orbiting craft crosses the equator from south to north is known as the orbit's ascending node; at the descending node, on the other side of the planet, the craft crosses from north to south.

28.5°

Apogee          Perigee

A craft in an elliptical orbit speeds up as it moves toward perigee—the point closest to Earth—because the planet's gravity and the ship's own orbital motion pull in the same general direction. Swinging back toward apogee—the point farthest from Earth—it moves against the pull and slows.

EQUATOR

# THE CRITICAL ANGLE

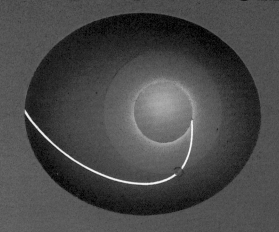

Earth's swaddling of atmosphere is both a barrier and a safety net for spacecraft. On the outbound journey, a ship seeking to go into orbit fights not only gravity but also atmospheric drag. On the return, however, the blanket of air helps to slow the craft for landing.

Technically, reentry begins at apogee, when the spacecraft is farthest from Earth and traveling slowest. By firing its retrorockets, the ship slows further, falling out of orbit into a long, curving earthward

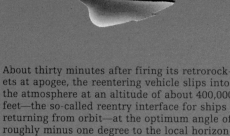

About thirty minutes after firing its retrorockets at apogee, the reentering vehicle slips into the atmosphere at an altitude of about 400,000 feet—the so-called reentry interface for ships returning from orbit—at the optimum angle of roughly minus one degree to the local horizontal. Exaggerated here for clarity, the angle is calculated from a theoretical horizontal line drawn perpendicular to a vertical line from the center of the Earth to the reentry point. To protect the ship and its crew from temperatures that can reach 3,000 degrees Fahrenheit, the craft is equipped with thermal shields.

trajectory *(diagram, far left)*. The trajectory has to be carefully calibrated: Too steep a dive and the spacecraft will become a meteor *(below)*; too shallow and it will skip off the top of the atmosphere and back into space along a different trajectory *(below, right)*.

The proper reentry angle allows the returning ship to do most of its braking in the tenuous reaches of the upper atmosphere while completing about half an orbit after retroburn at apogee. Though sparse, the supply of gaseous molecules in the upper atmosphere is sufficient to produce significant friction. As the speeding craft plows through the air, its movement agitates and compresses the molecules, surrounding the craft with a shock wave and converting the ship's kinetic energy to thermal energy. The resulting heat is so intense that it strips away electrons from the air molecules, producing a plasma of superhot particles.

A spacecraft that enters the atmosphere at too precipitous an angle generates so much friction that it burns up in the fierce heat from the surrounding gases. Even if the craft survived the fire, the rapid deceleration associated with such a steep plunge would produce a force so many times that of Earth's gravity that the crew would be crushed to death.

An attempted reentry at an angle shallower than minus one degree can also be disastrous. Like a flat stone skipping off the surface of a pond, the craft bounces off the top of the atmosphere. Depending on its velocity at the critical moment, and the precise angle of the attempt, the ship could be flung into an orbit that would make a safe landing impossible. A craft returning from the Moon could suffer a worse fate. The greater speed of a lunar reentry could produce a trajectory that would send ship and crew irretrievably into deep space. In certain circumstances, however, a returning ship could use atmospheric skipping to change its landing site *(page 72)*.

Tethered to the *Gemini 12* capsule by thirty feet of cord, Buzz Aldrin steps into the vacuum of space some 100 miles above Earth. The astronaut spent a total of five and a half hours outside the craft during the November 1966 mission.

inally comes the ten-second countdown and, clutching an emergency ejection ring between my legs in a death grip, I hold my breath and wait for ignition. There it is! A slight bump and I know we're on our way. I can see a tiny patch of blue sky out my window, but there is no sensation of speed until a thin layer of cloud approaches. Then—pow!—we burst through it faster than my eye can follow. We really are moving out. I begin to feel heavier . . . . At 50 seconds, we pass the altitude after which we can no longer use our ejection seats and I let go of the ring, my fingers tingling. Suddenly, as the power shifts to the second stage engine, I am flung forward in my straps . . . another lurch—and here we are, hanging in our harnesses, weightless at last. Out the window is the most amazing sight I have ever seen, a glorious panorama of sea and clouds stretching for a thousand miles in a glistening white light."

Such was Major Michael Collins's description of his first voyage into space, a journey made aboard *Gemini 10* on July 18, 1966. "These are not ejection seats," he wrote, "but thrones facing out on the Universe—and we are wealthier than kings." His exhilaration and awe were reactions common to the more than 200 spacefarers who would rocket into orbit in the course of three decades. The account is typical also in that, remarkably, it reveals the merest hint of fear, despite mortal dangers of many kinds.

Although earlier visionaries had solved many of the basic challenges of ballistics and propulsion and had put small payloads into orbit, lofting a human being into space required a whole new level of technology. To survive a journey into the lethal vacuum beyond Earth's atmosphere, astronauts must carry much of their complex environment with them. They must be supplied with oxygen and relieved of their poisonous exhalations of carbon dioxide: These two gases have to circulate at a rate that maintains the spacecraft's atmospheric pressure at a minimum of five pounds per square inch (psi). (Normal, sea-level pressure is fourteen psi, but astronauts can breathe at five psi if the gas is pure oxygen.) The cabin must be hermetically sealed; by the time the craft passes 63,000 feet, air pressure outside is so low that bubbles would form in the blood and water vapor would fill the lungs of an unprotected human. Furthermore, a ship in the near vacuum of space is no longer shielded by an atmosphere from temperature extremes. In sunlight, an uninsulated craft could quickly heat up to 200 degrees Fahrenheit; in shade, it could drop to minus 150. Even with the proper insulation, crew members face another hazard: Heat generated by their own bodies and by cabin equipment can raise temperatures inside the cabin to unpleasant levels.

The universe beyond Earth's atmosphere is rife with other perils as well. The so-called Van Allen belts—bands of high-energy protons and electrons that lie along the Earth's magnetic field lines—contain enough radiation to kill a suited-up astronaut, as do streams of similar particles occasionally emitted by the Sun. Even more dangerous are meteoroids—rocky matter, left over from the formation of the Solar System, that whizzes through the void at velocities approaching 50,000 miles per hour. A metal skin one-eighth of an inch thick can be penetrated by rock fragments no bigger than a pebble. To date, spacecraft designers have not built in radiation or meteoroid shields, believing that such dangers are either avoidable or rare.

The reason engineers have taken this gamble is that with every protective shield and life-support system, the spacecraft gains mass and therefore demands more propulsive energy. And that is only the beginning of the weight problem. Radio communications—and all the equipment they entail—are needed to keep the craft in touch with the ground crew controlling the flight. Safe landings necessitate an array of parachutes, escape systems, and padding. Pile on enough food and water for a journey of only a few days, and a vehicle capable of providing the barest minimum of these essentials weighs about 3,000 pounds. To lift something that heavy out of the atmosphere requires a propulsion system capable of at least 78,000 pounds of thrust.

Less than a month after *Sputnik 1* appeared in the skies, Soviet engineers took the first run at this daunting challenge. Although the passenger aboard *Sputnik 2* was not human and the craft not designed for reentry and landing, the mission tested many of the other elements needed for human flight. A brown and white mongrel named Laika was lifted into orbit on November 3, 1957, and circled Earth in the Volkswagen-size capsule for eight days. On the eighth day, when her oxygen was about to run out, Laika ate a scheduled ration of food that had been pretreated to euthanize her painlessly. *Sputnik 2* eventually broke up on reentering the atmosphere.

**THE RACE TO GET READY**

Obviously the Soviets had human flight in the works, a prospect that galvanized U.S. policymakers. Within five days of *Sputnik 2*, Congress launched an official inquiry into U.S. space efforts, and in the summer of 1958, President Dwight Eisenhower signed an act creating the civilian National Aeronautics and Space Administration.

Within six months, a hastily assembled team of twenty scientists had framed the U.S. space program for the next ten years. In 1959, NASA persuaded the government to transfer part of the army's missile program to the agency—along with the program's presiding genius, Wernher von Braun, who was developing the massive Saturn rocket later used in Apollo missions.

Yet as rapidly as NASA got itself in gear, the Soviets dominated the early years of the space race. Aeronautic mastermind Sergei Korolyov, known outside the USSR only by the shadowy title Chief Constructor, achieved one spectacular after another. In addition to a series of unmanned lunar probes, the Soviets launched *Sputnik 3* in 1958; the satellite was more than two and a half times heavier than *Sputnik 2* and stayed aloft for two years.

As these successes accumulated, the Soviet Union seemed destined to prevail in planetary exploration as well. Then they missed a beat. Many space watchers expected the Soviets to launch a high-profile, unpiloted mission to Mars in conjunction with Premier Nikita Khrushchev's visit to the United Nations in October 1960. But Khrushchev came and went, and beyond his shoe-banging tirades nothing happened. A few weeks later, the Soviets announced that Field Marshal Mitrofan Nedelin, commander in chief of Strategic Rocket Forces, had died in an airplane crash on October 25.

In the years since, several sources, including Khrushchev himself in his posthumous memoirs, have reported that there was more to Nedelin's demise than a mere plane crash. According to the accounts, the commander, driven by Khrushchev's desire for a United Nations propaganda coup, decided to launch the Mars-bound rocket before it had been thoroughly checked out. The countdown reached zero just after sunset on a day in late October. An ominous silence followed. The rocket, containing close to a million pounds of explosive fuel, remained immobile as observers waited nervously in the dark. Elementary precaution ruled out a close inspection, but finally Nedelin's impatience won out, and he ordered the flight engineers to examine the rocket. They had been on and around it for more than an hour when the booster detonated in a titanic blast that must have been heard a hundred miles away. In a few nightmare instants, Nedelin and anywhere from forty to several hundred other members of the Soviet rocket brain trust were incinerated.

Yet the disaster hardly slowed the headlong pace of Soviet experimentation. *Korabl Sputnik*s (the name meant "satellite ship") *2, 3,* and *4,* put up in 1960 and early 1961, carried a varied menagerie of dogs, mice, rats, and insects. On one flight, a human mannequin was rigged to play a tape recording of the Piatnitsky Russian choir into the radio so controllers could test the equipment. U.S. scientists could only conclude that the Soviets were on the verge of launching a piloted mission.

## ROCKETS WITH AND WITHOUT WINGS

Though racing to catch up with their Soviet counterparts, engineers in the United States were not completely inexperienced in the business of sending humans into space. High above the desiccated lake beds of Edwards Air Force Base in California, a band of daredevil pilots was already testing the outer limits of manned flight. Satellites such as the Sputniks were one-time cannon shots, but the rocket-powered X-15s flown by these men were authentic self-propelled and self-guided aircraft. Long, sleek metal cigars with stubby wings set well back on their fuselages, X-15s had their roots in the ultrafast jets developed near the end of World War II. Strapped to the bottom of a B-52 bomber, the airplane was carried up to 45,000 feet, the B-52's altitude limit, and cut loose. As its powerful engine ignited, burning 20,000 pounds of liquid fuel in eighty seconds, the plane shot even higher, accelerating to more than 3,000 miles per hour—sometimes much more. On one flight, the rocket-plane, painted with a nickel-steel alloy to keep it from scorching, achieved a speed of 4,520 miles per hour and an altitude of sixty-seven miles. In writing about the space program later, Wernher von Braun called the X-15 "the closest thing to a winged spacecraft that has ever been built."

The missions yielded invaluable information on the behavior of both aircraft and humans in suborbital space. At high altitudes, the lack of airflow rendered aeronautic control devices such as rudders and ailerons useless; in addition to such controls, X-15s had spacecraftlike attitude-control thrusters, small hydrogen peroxide jets on the nose and wingtips to aid in steering. Bundled up in proto-spacesuits, the pilots proved that humans could survive high acceleration and at least short bouts of weightlessness.

## PROJECT MERCURY

Despite the ongoing success of the X-15, NASA engineers turned away from the traditional winged design in 1958. Their objectives were few and deceptively simple: to launch a vehicle carrying a human being into at least one orbit around Earth, to recover the craft and its payload in one piece, and to scrutinize the effects of this alien environment on the pilot. The program, named Mercury, amounted to a crash course in space flight that would form the bedrock of all later U.S. piloted space missions.

Project Mercury's designers, led by NASA's innovative aerodynamics expert Maxime Faget, concluded that wings and wheels were heavy anachronisms in the space age. With the powerful rockets available by the 1950s, a spacecraft could simply be lifted, piggyback, into space. It did not have to be sleek: In fact, Faget felt that the optimal design was a cone-shaped vehicle, able to slow its reentry by controlled atmospheric friction. Traveling at high speeds, the blunt end would generate a superheated bow shock wave ahead of it, while the craft itself would stay relatively cool.

As designed, the Mercury spacecraft would rest on top of von Braun's Redstone booster rocket. Only six feet in diameter, the unlovely vehicle managed to provide just enough space for one astronaut and 40,000 separate components. On-board instruments passed a constant stream of information to ground control on details ranging from the capsule's temperature to the astronaut's heart rate.

Riding Redstone's fireball, Mercury had no more control over its trajectory than a snub-nosed bullet. Once in orbit, the capsule would travel automatically along a preset flight path, although a pilot could make small adjustments using manual controls. A mechanism called the Attitude Stabilization and Control System (ASCS) determined the capsule's pitch (up and down movement), yaw (sideways movement), and roll (spin along its own axis), fine-tuning these alignments with bursts from small jets on the capsule's hull.

The ASCS allowed an astronaut to orient the vehicle for experiments and for its return to Earth.

For reentry, retrorockets would slow the craft until gravity began to pull it back toward the ground. Dropping bottom first into the atmosphere, the conelike capsule would be protected from the 3,000-degree temperatures of the bow shock wave by a heat shield attached to its base. Mercury's final descent would slow to a leisurely twenty miles per hour under a sixty-three-foot-wide parachute; at the last minute, an air bag under the capsule would inflate to cushion the blow of an ocean landing.

## THE AMERICAN SEVEN

While Faget designed the perfect vehicle, NASA searched for perfect passengers. President Eisenhower narrowed the field considerably by decreeing that all candidates be military test pilots, since they had already demonstrated cool heads and fast reflexes. NASA refined the requirements to young (under forty) college graduates, no taller than five feet, eleven inches (so they could squeeze into the tiny capsule), and in superb physical condition.

A group of 110 passed the first screening early in 1959, and 32 of the first 69 interviewed volunteered immediately—more than enough to satisfy NASA's needs. At the Aeromedical Laboratory of Wright-Patterson Air Force Base in Ohio, flight doctors separated the superhuman from the merely stalwart. Tests verged on the diabolical. In order to gauge their reactions to physical shocks, for example, the candidates were shut for two hours in a room heated to 135 degrees.

Eighteen survivors then submitted to a week of psychological scrutiny intended to select the seven pilots who would most likely remain congenial

# A DECADE OF ACHIEVEMENTS

From 1957 to 1966, astronauts and cosmonauts—and a few animals before them—scored some two dozen space-flight firsts. Spurred by the Soviet-American race, the rapid-fire series of achievements, some of which are noted here and on the following pages, laid much of the groundwork for later travel to the Moon. The Soviets were first to prove that human beings could orbit the Earth and "walk" in space outside their vehicles; they also fielded the first multiperson flight crew. Then, in the mid-1960s, the United States edged ahead with the first rendezvous and docking of two orbiting craft—a key requirement for NASA's planned Moon landing. A few years later, the first humans to set foot on another world would use similar techniques to rejoin their ship and return to Earth.

**1957** A mongrel dog named Laika was the first animal in space, orbiting Earth for a week aboard *Sputnik 2* as Soviet scientists monitored her physical reactions. On the eighth day, pretreated food put Laika painlessly to sleep.

**1959** Female monkeys Able *(below, left)* and Baker *(below, right)* were the first animals to return alive from space, surviving a fifteen-minute flight to a height of about 300 miles in the nose cone of an American intermediate-range ballistic missile.

in close quarters over long periods of time. These turned out to be John Glenn, LeRoy Gordon Cooper, Malcolm Scott Carpenter, Walter Schirra, Virgil "Gus" Grissom, Alan Shepard, Jr., and Donald "Deke" Slayton (later sidelined because of a minor heart ailment). All were husbands and fathers, all small-town boys who had grown into world-class test pilots.

The first astronauts' training focused as much on the physical experience of space flight as on the operation of the spacecraft itself. Space physicians, worried about the tremendous forces of acceleration the men would feel, had already devised several frighteningly authentic experiments. In one, they shot a brave volunteer in a rocket-powered sled at almost 1,000 feet per second along a 3,500-foot track that ran into an oversize bathtub, learning that humans could survive forty g's, or forty times normal Earth gravity. However, NASA decided that twelve g's was the most astronauts could be expected to withstand and still perform their duties.

NASA flight doctors were also concerned about the reaction of their patients' metabolisms to weightlessness. They did not know how much the human body might rely on gravity for its functions. Would internal organs continue to work? Would the astronauts' lauded reflexes prove as lightning quick without the normal orientation of gravity? The physicians tried to answer these questions by putting the seven men through their paces in jets flown along ultrahigh, ultrasteep arcs. At the top of these stratospheric loops, they experienced twenty precious seconds of zero gravity, during which they moved, ate, or drank at a frantic pace—proving that weightlessness was safe, at least in small doses.

Between exercises, the seven made sure that Mercury's design conformed to their own expectations, using their collective clout to push through several key modifications. The revised design included more legroom, a window large enough to see out of, and an escape hatch that could be blown away, allowing the crew to escape if the downed capsule began to sink in the ocean.

**1961** Yuri Gagarin, the first human being in space, orbited the Earth on April 12. His flight aboard *Vostok 1* lasted one hour and forty-eight minutes and ended with a safe landing in a Russian field.

**1961** The first American in space, Alan Shepard, made a fifteen-minute suborbital flight in *Freedom 7 (right)* on May 5. During the mission, Shepard pioneered the use of on-board instruments to control a craft in space.

**1962** On February 20, John Glenn circled the Earth three times in *Friendship 7,* becoming the first American in orbit. Here, he relaxes aboard the USS *Noa* soon after his flight, which lasted four hours and fifty-five minutes.

The foremost concern of the Mercury astronauts, however, was not safety so much as the amount of control they could exert over the spacecraft in flight. They were test pilots, after all, and none of them cared to ride as mere passengers—"spam in a can," as they called it. It was bad enough that NASA intended to put a chimpanzee into space ahead of them. Under pressure, NASA allowed a compromise system called fly-by-wire, which enabled the astronauts to bypass the ASCS to control the capsule's attitude thrusters directly. By moving a joystick back and forth, side to side, or in a corkscrew motion, pilots could change the ship's pitch, roll, and yaw.

On February 21, 1961, after several successful test flights (including one that had been "piloted" by the chimpanzee, named Ham), NASA officials announced to the world their intention to put the first man in space. On April 12, they ate their words when cosmonaut Yuri Gagarin, a twenty-seven-year-old lieutenant in the Soviet Air Force, climbed into a seven-and-a-half-foot capsule atop a modified ballistic missile and calmly blasted off into a complete orbit around the Earth.

## FIRST STEPS INTO SPACE

A true working-class hero, born 100 miles to the west of Moscow in the small town of Gzhatsk, Yuri Gagarin—like the Mercury seven—was a crackerjack jet pilot who had passed a fearful battery of physical and psychological tests to win the honor of being the first human in space. Short (he had to sit on a cushion when flying jets), stocky, and stubborn, Gagarin seemed entirely undaunted by the

**1963** Valentina Tereshkova, the first woman in space, orbited Earth aboard *Vostok 6* for two days and twenty-three hours. During her mission, *Vostok 5* approached within three miles, an early rendezvous maneuver.

**1964** The first space team was the Soviet trio of *(clockwise from top)* pilot Vladimir Komarov, physician Boris Yegorov, and physicist Konstantin Feoktistov, who spent just over a day in orbit aboard *Voskhod 1*.

**1965** Cosmonaut Alexei Leonov took the first walk in space, remaining outside the *Voskhod 2* capsule for eighteen minutes.

magnitude or the risk of his endeavor, shouting "Off we go!" as his rocket left the Russian steppe at the Baikonur Cosmodrome in Kazakhstan.

The cosmonaut's capsule, named *Vostok 1*, was launched atop a powerful A-1 rocket, capable of boosting far more weight into orbit than any of its American counterparts. As a consequence, Gagarin's craft was much larger than the first Mercury capsules, with a cabin that was roomy by space standards. The extra room allowed the cosmonaut to breathe standard air at sea-level pressure, unlike Mercury astronauts, who would save space with a simpler apparatus that fed in pure oxygen. After *Vostok 1* reached orbit, Gagarin made one complete global revolution—25,400 miles—in an hour and forty-eight minutes. During his journey, he listened to Tchaikovsky, and as he fell back to Earth he lustily sang a patriotic song.

In contrast to NASA's plans for sea landings, the Soviets had chosen to recover their spacecraft on dry ground, in part because their vehicles were large enough to absorb the additional shock. The decision also reflected the Soviet penchant for secrecy, which led them to forgo a more public splashdown in open water in favor of sites hidden in the interior of the USSR. *Vostok 1* thus floated more or less gently to the surface under a canopy of giant parachutes, touching down in the rural community of Smelovaka before the wide eyes of two speechless farmers.

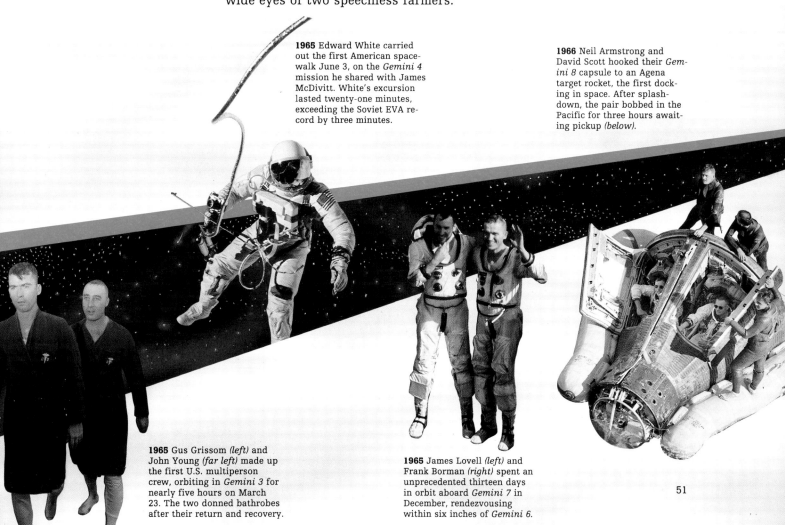

**1965** Edward White carried out the first American spacewalk June 3, on the *Gemini 4* mission he shared with James McDivitt. White's excursion lasted twenty-one minutes, exceeding the Soviet EVA record by three minutes.

**1966** Neil Armstrong and David Scott hooked their *Gemini 8* capsule to an Agena target rocket, the first docking in space. After splashdown, the pair bobbed in the Pacific for three hours awaiting pickup *(below)*.

**1965** Gus Grissom *(left)* and John Young *(far left)* made up the first U.S. multiperson crew, orbiting in *Gemini 3* for nearly five hours on March 23. The two donned bathrobes after their return and recovery.

**1965** James Lovell *(left)* and Frank Borman *(right)* spent an unprecedented thirteen days in orbit aboard *Gemini 7* in December, rendezvousing within six inches of *Gemini 6*.

51

Gagarin's achievement, combined with his charm, earned him acclaim both within the Soviet Union and elsewhere. In the years to come, he was showered with awards by countries around the world and even had a crater on the Moon named after him. But he never flew another mission. For unknown reasons, the Soviets did not send him into space again. In 1968, to national sorrow, the cosmonaut was killed in a plane crash.

Meanwhile, the Americans were playing catch-up. Three and a half weeks after Gagarin's flight, at 9:34 a.m. on May 5, 1961, commander Alan Shepard felt the Redstone engines thunder beneath him as *Freedom 7* lifted off from the pad at Cape Canaveral. What began as a smooth ascent grew a lot rougher about a minute after liftoff—so rough that Shepard wondered if the rocket would rattle to pieces. The taciturn New Hampshire Yankee kept this opinion to himself, fearing that mission control might abort the flight, and to his relief the shaking passed when the capsule separated from the rocket and slipped into space. As he entered zero gravity, Shepard tried operating the capsule on fly-by-wire. It worked so well that he turned off the ASCS at the peak of the trajectory for some fine-tuning before the retrorockets fired and the ship began to fall earthward. The descent was fast—a little too fast—and Shepard had a few more anxious moments before his chute deployed in perfect form, gently depositing *Freedom 7* in the ocean.

The fifteen-minute flight—deemed "just a pleasant ride" by its laconic pilot—catapulted the United States into the space race. Only three weeks after Shepard's triumph, President John Kennedy upped the ante, declaring that the U.S. intended to put an American on the Moon by the end of the decade.

Within two months of Kennedy's announcement, the second Mercury flight went up. Gus Grissom, a burly, thoughtful man, took off on an almost identical mission on July 21. The day was clear, and the flight went perfectly—until Grissom splashed down. While he was waiting for the recovery team, the escape hatch on his capsule blew off prematurely, and the Atlantic poured in. Grissom managed to escape from the foundering capsule, only to be pressed into the water by the downdraft of a rescue helicopter. Piling insult on injury, the chopper concentrated on snapping pictures of the hapless astronaut, while another unsuccessfully tried to save the waterlogged capsule. Yet another copter passed him by; finally a fourth rescued the sodden spacefarer.

Despite the loss of the capsule, the mission was judged a success. American jubilation was cut short, however, when the Soviets again moved forward on August 6, 1961, launching Captain Gherman Titov into orbit in *Vostok 2*. This time, the craft flew for slightly more than a day, covering a total of 436,656 miles in seventeen orbits. Titov's face was broadcast on televisions around the world, as was a recording of his voice from Vostok's cockpit shouting "I am Eagle! I am Eagle!" The exuberant twenty-five-year-old achieved many mundane space firsts as well. He was the first cosmonaut to get spacesick and the first to sleep in space, although he had to keep his weightless arms restrained under a safety belt to prevent them from floating around and waking him up.

The stakes had been raised again: Before the United States had put even one man in orbit, the Soviets had lofted two, and the cosmonauts were starting to log significant time in space during each flight. Merely sending an astronaut up and down again in a grand arc would no longer cut it. NASA officials decided they, too, were ready to put an astronaut into Earth orbit. The agency scrubbed a planned third suborbital mission and readied its pilot, John Glenn, for a flight three orbits in duration.

If Glenn was lucky to have snagged the assignment, NASA was equally fortunate to have had him piloting its boldest mission to date. Already a flying hero, having set a transcontinental airspeed record in 1957, Glenn was also the most self-conscious role model of the bunch. As one of his friends said, "John tries to behave as if every impressionable youngster in the country were watching him every moment of the day."

## THE FLIGHT OF FRIENDSHIP 7

To boost the spacecraft to the ninety-seven-mile orbiting height, NASA decided to employ an Atlas 109D rocket, America's first intercontinental ballistic missile (ICBM), built to deliver thermonuclear warheads a third of the way around the world. Glenn's Mercury vehicle, named *Friendship 7*, lifted off on top of the ten-story-high Atlas on February 20, 1962, watched by a television audience estimated at 100 million in the United States alone. Once in orbit, Glenn reveled in his first prolonged experience of weightlessness. "You feel absolutely free," he later wrote. "A person could probably become addicted to it without any trouble."

On Glenn's second orbit, the mission ran into what looked like serious trouble. A sensor on the spacecraft signaled to ground control that Friendship's heat shield was loose. Held in place by locks, the shield was in fact designed to come loose—but not until the final stage of descent, when it was supposed to drop down and release a landing bag to take up some of the shock. If the shield came off too early, the ship would burn up on reentry. NASA figured Glenn's best strategy was to use the capsule's retropack—the assembly of retrorockets needed for reentry—as an extra lock. Strapped outside the heat shield and directly attached to the capsule, the retropack was normally jettisoned after the retrorockets were fired, so that the heat shield could dissipate heat most efficiently. Although leaving the pack on would cause some problems, ground controllers decided they were much the lesser of evils.

Once the retros had fired and *Friendship 7* was irrevocably committed to reentry, Al Shepard broke the news to his friend. "We are not sure whether or not your landing bag has deployed," he said. "We feel it is possible to reenter with the retropackage on." "Roger," replied Glenn, "understand." His voice was calm, but according to the pulse sensors attached to his chest, his heart rate shot up to 132. As if the heat shield was not enough to worry about, the ASCS was malfunctioning, and Glenn had to work feverishly to keep the capsule pointed downward at a perfect angle so that the unprotected sides of his capsule would not melt in the searing frictional heat. As *Friendship 7* entered the atmosphere, the retropack began to burn. Big chunks of it flew past Glenn's window in fireballs. "Those few moments," he later wrote, "ticked off inside the capsule like days on a calendar."

As it turned out, the craft's heat shield had actually been intact during reentry; the sensor issuing the distress signal had been faulty. Glenn's capsule splashed down forty miles from its recovery ship, having completed its three trips around the globe in four hours and fifty-five minutes. President Kennedy flew to the Cape to meet him, and a few days later Glenn was given a triumphal New York City ticker-tape parade.

Two more Mercury missions followed *Friendship 7* in the same year. In May, Scott Carpenter flew another three-orbit flight. The enthusiastic astronaut made liberal use of the manual controls—perhaps too liberal. He splashed down depleted of fuel and 250 miles beyond his target point. Wally Schirra flew a mission almost twice as long—five and three-quarters orbits— in October. As the novelty of Americans in orbit wore off, however, these missions drew less and less attention. Schirra's flight was almost perfect, and yet one of the witnesses at the launch pad called it "dullsville." This business-as-usual attitude seemed to infect even the astronauts. While waiting for liftoff during the last Mercury mission on May 15, 1963, Gordon Cooper fell asleep during the countdown. The youngest of the Mercury pilots and a native of Shawnee, Oklahoma, "Gordo" remained utterly relaxed throughout the flight's twenty-one and three-quarters orbits, even squeezing in a seven-hour snooze. Nevertheless, he accomplished a great deal—deploying a tethered balloon and a flashing beacon, and photographing Earth in detail fine enough to pick out the wake of a boat on the Nile River 100 miles below.

**AND NOW FOR PHASE TWO**

With Cooper's ride, the Mercury portion of the space program was over. In the Soviet Union, the Vostok series was also coming to an end. During 1962, *Vostok*s *3* and *4* were launched one day apart, and the two capsules passed within a few miles of each other in orbit. In June of 1963, the Soviets took another symbolic step forward when Valentina Tereshkova became the first woman in space. The former cotton-mill worker and amateur parachutist lifted off in *Vostok 6*, two days behind *Vostok 5*. Like the earlier Vostok pair, the two spacecraft passed within a few miles of each other during their tandem journeys. Not only did Tereshkova perform admirably, but she was also an incomparable political asset. "As you can see," she proclaimed later, "on earth, at sea, and in the sky, Soviet women are the equal of men."

In three hectic years, both the Mercury and the Vostok programs proved that piloted space flight was not just the fantasy of science-fiction writers. Humans could indeed survive the bruising pressures of launch and reentry, and they could breathe and function intelligently in space, despite the effects of weightlessness. Soviet and American scientists were now ready for even more daring ventures.

In the mid-1960s, mission planners on both sides of the globe began to prepare the way for voyages to space stations and the Moon. The next round of flights would include not only multiperson crews but extravehicular activities (EVAs) and dockings as well. All this called for much more complex engineering than the Vostok or Mercury programs had employed.

As they had before, the Soviets took the lead. On October 12, 1964, the Russian spacecraft *Voskhod 1* made history by placing three men in orbit for twenty-four hours. NASA officials were impressed, but a few of them wondered why the craft stayed aloft for only one day, when rumors had led them to expect a week-long mission.

Over time, it was revealed that Khrushchev had again ordered his engineers to move quickly before the United States could launch a new generation of piloted flights. Faced with an impossible deadline, chief designer Korolyov simply stripped the original one-person Vostok down to its barest essentials to make room for three passengers. Those three—cosmonauts Vladimir Komarov, Konstantin Feoktistov, and Boris Yegorov—may have been the most courageous people ever to enter a spacecraft, for in the process of conversion Korolyov removed several critical safety features. Not only was the capsule deprived of its ejection seats and reserve parachute, but also the crew could wear no spacesuits, because there was no room for them; the cosmonauts had to sit sideways as it was. Under these conditions, keeping the mission short seemed only prudent.

In a twist of fate, Khrushchev himself was ousted from office while the flight was under way. But the new Soviet leadership was no less intent on rapid progress. *Voskhod 2*, launched five months later in March 1965, featured another innovation: an EVA—or spacewalk, in general parlance—by cosmonaut Alexei Leonov. Despite some difficulties with the maneuver, the feat showed that the Soviets were still one giant step ahead of the Americans.

*Voskhod 2* was the last piloted Soviet flight for two years. Many space scientists have since concluded that the Voskhod missions, while impressive, may have cost the Soviets the Moon by bleeding resources away from more necessary technological developments. To make matters worse, Korolyov—guiding spirit of the Soviet space program—died in early 1966.

## ACCELERATING TOWARD THE MOON

In the United States, too, space engineers were feeling the pressure. When NASA first planned the Mercury missions, it intended to follow them with a long series of Apollo flights, a strategy that would have put Americans on the Moon in the mid-1970s. But when President Kennedy decided that the lunar landing should take place in the late 1960s, NASA had to drastically accelerate its schedule. Although a rocket powerful enough to carry humans to the Moon was still years away, spacecraft launched with existing rockets could allow NASA to perfect most other aspects of such a mission. With this shortcut in Apollo's development cycle, the new deadline was feasible.

NASA wanted eventually to launch astronauts into lunar orbit in a vehicle carrying a separate landing module. The module would descend from the

# THE WATCH FROM EARTH

Much as a toddler learns to walk by being passed from one outstretched hand to the next, twentieth-century pioneers in the infancy of space flight—whether orbiting the Earth or en route to the Moon—are, in effect, handed off from one earthbound tracking station to another as their journey and Earth's own rotation bring their craft up over the horizon *(below)*. As described on the following pages, the stations are strategically placed to provide maximum coverage for missions in low Earth orbit and farther out, relaying all information to a centralized mission control.

At each tracking site, an electronically governed radio antenna picks up signals from the craft and transmits signals to it, automatically forwarding data from this exchange to mission control. The signals between craft and tracking station are transmitted on prearranged frequencies according to a formula that allows a computer at mission control to almost instantaneously determine the ship's distance, celestial position, and velocity. Mission controllers compare that information to the flight plan and decide whether minor course corrections, or more drastic action, are needed.

Ship-to-station links are also used for television and voice communications and for relaying scientific and engineering data. Because of interference from a variety of sources, all space tracking and communications take place on radio frequencies between 30 gigahertz (GHz) and 100 megahertz (MHz). (One hertz equals one wave cycle per second; a megahertz is one million hertz, a gigahertz one billion.)

**Determining distance.** A ground station finds the distance, or range, to a spacecraft by clocking the time it takes radio signals, traveling at the speed of light, to go from ground to ship and back. The station sends a signal and waits for an acknowledgment sent by an on-board device called a transponder. Multiplying half of the round-trip time by the speed of light (186,000 miles per second) produces the one-way distance. Range measurements are accurate to within 100 feet. The same signal exchange is used to determine speed and position.

355°

350°

345°

**Fixing position.** As with any celestial body, a spacecraft's location is determined by two coordinates: its elevation in degrees above the horizon and its azimuth in degrees along the horizon from true north. The station points its antenna toward the position in the sky where the craft is expected to appear over the horizon, then sweeps the area until it homes in on the ship's signals. When the signals are strongest, special devices on the antenna encode the antenna's direction and tilt, which in turn give measurements of azimuth and elevation accurate to within a hundredth of a degree.

Water Vapor

10 GHz  |  1 GHz  |  100 MHz  |  10 MHz

Human-Generated Noise

As illustrated here, various kinds of interference limit the portion of the radio spectrum available for space-flight tracking and communications. Atmospheric water vapor starts to absorb radio at frequencies above 10 GHz and becomes a severe problem at higher frequencies. Below 100 MHz, human-generated, cosmic, and solar noise produce static, and the ionosphere reflects radio waves downward (arrows). Early missions used ultrahigh frequency, or UHF (300 MHz), for voice communications, and C-band (5 GHz) and S-band (2 GHz) for tracking. Beginning with the Apollo flights, S-band has been employed for both voice communications and tracking; K-band (15 GHz) transmits television signals and high-speed data on shuttle flights.

**Measuring speed.** To verify that a spacecraft is maintaining its proper course and velocity, tracking stations make use of the phenomenon known as the Doppler effect. Just as the pitch of a moving siren changes from high to low as it passes, the frequency of a signal depends on the relative movement of the emitter and the receiver. The signal transmitted by the ground station is on a predetermined frequency. When the ship's transponder receives it, it sends back a signal on a frequency that is different by a predetermined amount. Because of the Doppler effect, the frequencies of the signals received at either end vary from the arranged emissions. Mission control's computer notes the variance, calculates the Doppler effect, and finds the ship's speed.

N
|
0°

5°

Like ripples spreading outward from a pebble dropped in a pond, radio signals propagate in all directions from a source, growing weaker with distance. The dish-shaped reflector of a radio antenna concentrates the energy of the signals into a beam, rather like a flashlight; the resulting increase in signal strength is known as gain. Engineers trying to maintain maximum contact with space missions must balance the size and gain of the emitting antenna against the distance the signal must travel and the size and gain of the receiving antenna.

Because the radio devices aboard most spacecraft are small, transmitting at less than 10 watts, they need powerful partners on the ground. Tracking stations use huge dishes, from 30 to 210 feet across, to gather the faint radio waves lapping Earth from on high. To communicate with a small moving target, the ground antennas in turn transmit at 2,000 watts (up to 10,000 watts) and focus the signals into a beam no wider than one degree of arc.

The amount of coverage provided by a ground network is determined by three factors. One is that antennas need a line-of-sight view of the spacecraft: They cannot track craft below the horizon, or even—for various reasons—all the way to the horizon. A dish thus has a maximum zone of coverage of about 170 degrees of the sky *(opposite, top)*. The second factor is the craft's altitude: The higher up the ship, the longer a ground antenna can keep it in range. Finally, because Earth rotates under a spacecraft and the craft's orbit is inclined to the equator, a ship will not pass over every ground station on every orbit, a problem for craft at low altitudes *(opposite, middle)*. In 1983, NASA turned to a satellite-based system *(opposite, bottom)*, which can keep low-orbiting craft under almost constant surveillance.

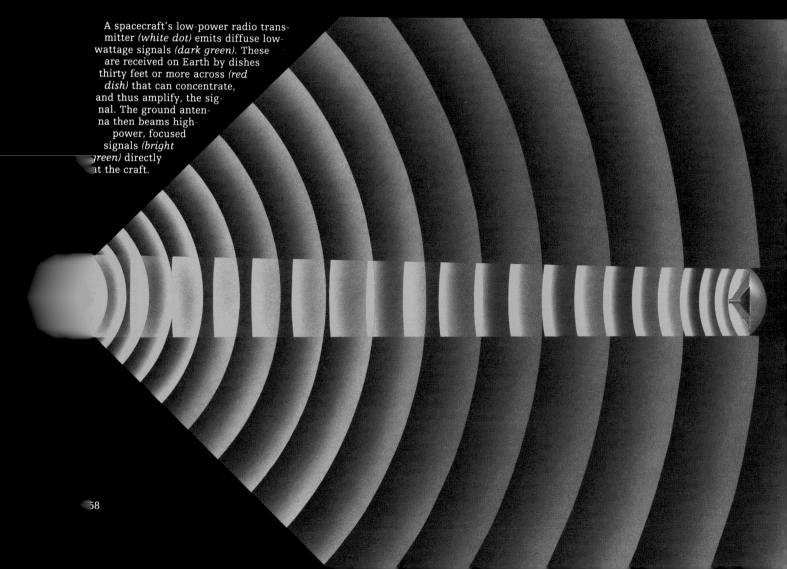

A spacecraft's low-power radio transmitter *(white dot)* emits diffuse low-wattage signals *(dark green)*. These are received on Earth by dishes thirty feet or more across *(red dish)* that can concentrate, and thus amplify, the signal. The ground antenna then beams high-power, focused signals *(bright green)* directly at the craft.

A ground network with three stations, each with a maximum 170-degree zone of coverage *(blue area)*, can provide nearly continuous coverage for distant craft *(outer circle)*, leaving only narrow gaps *(black triangles)* when the ship passes out of view. However, for craft such as the shuttle, which orbits at altitudes ranging from 100 to 160 miles *(inner circle)*, a three-station network leaves large portions of the orbit uncovered.

To provide more continuous coverage for spacecraft in low Earth orbit, a ground network must consist of many more stations. Early piloted missions, for example, employed about two dozen stations. (For simplicity, only six are shown at right.) Even at that, the combined effects of Earth's rotation and the inclined plane of the shuttle's orbit meant that the ship was in contact with mission control for only fifteen minutes of every ninety-minute circuit.

With the Tracking and Data Relay Satellite System *(TDRSS,* pronounced "tea-dress"), NASA needs ground stations only during launch and landing. The system consists of a ground antenna in New Mexico *(red dish)* and two satellites *(white dots)* in geosynchronous orbit, 22,300 miles above the Earth. At that altitude, the satellites complete one orbit in exactly the time it takes Earth to rotate once on its axis and are effectively stationary with respect to points on the Earth. Except for a narrow zone of exclusion less than 200 miles wide over the Indian Ocean, they provide almost total coverage for any craft in low Earth orbit. Above 750 miles, TDRSS coverage is continuous.

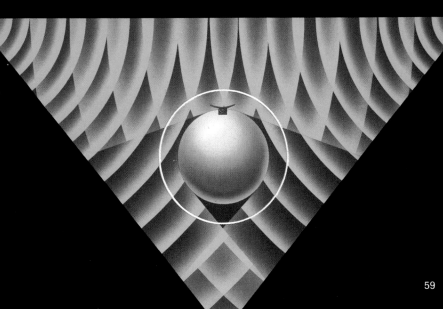

59

orbiter to the Moon's surface and at the end of the mission would reconnect with the orbiter for the trip home. America's pre-Apollo flights would therefore concentrate on such maneuvers as joining two spacecraft in orbit, extravehicular activities, and precisely controlled reentries and landings. Because these exercises required two astronauts, the program was named Gemini, after the zodiac sign of the twins.

The new program began inauspiciously, eighteen months behind schedule. But on March 23, 1965, a ninety-foot Titan II rocket, three times as powerful as the Atlases that had lifted Glenn and Schirra, boosted Gus Grissom and John Young in *Gemini 3* into orbit. During their near-perfect flight, the astronauts demonstrated Gemini's advantages over the Mercury capsule. For example, Gemini's new sixteen-rocket attitude and maneuvering system enabled the crew to change the altitude of the craft's orbit. Like many other modifications, the system gave more control to the pilots and brought the craft much closer to being a true spaceship.

The crew of *Gemini 4* was able to show off yet another instance of NASA's new technology. On June 3, 1965, after a clean liftoff and two orbits around the Earth, commander James McDivitt and his copilot Edward White made a final check to ensure that their spacesuits were sealed and pressurized. Then they slowly bled the oxygen out of their cabin until it was as airless as the void surrounding it. White opened the hatch above his head, stood up in his seat, and floated into open space, tethered by a twenty-five-foot, gold-coated umbilical cord. "This is the greatest experience," he cried. "It's just tremendous." The astronaut happily propelled himself around with an air gun, knocking the capsule out of alignment every time he bumped into it and all but ignoring orders by McDivitt to reenter the craft.

The fun ended when White finally climbed back in after twenty-one minutes: The hatch would not close properly, and the astronauts had to repair it by hand. Then, not wanting to take the risk of reopening the faulty door, they were forced to share the remainder of their voyage with a growing pile of garbage. By the time it landed on June 7, *Gemini 4* had completed sixty-two orbits, and America was within range of longer Soviet flights. Moreover, White's EVA, though somewhat uncontrolled, was the longest on record.

With two flights under its belt, the Gemini series moved on to longer missions to determine whether astronauts could endure a trip protracted enough to travel to the Moon and back. Two months after *Gemini 4,* Mercury veteran Gordon Cooper and his partner Charles "Pete" Conrad, Jr., set a new endurance record in *Gemini 5* of 120 orbits and nearly eight weightless days, with no apparent ill effects.

Mission planners pressed on to the next milestone: the docking of two vehicles in space. The project started in disaster. A thirty-two-foot-long Agena rocket booster, designed to act as a docking target, was launched on October 25, forty-eight days in advance of the scheduled launch of *Gemini 6.* But it crumbled to pieces before it reached orbit. Astronauts Wally Schirra and Thomas Stafford settled back for a long postponement.

Three days later, President Lyndon Johnson made the surprise announcement that *Gemini 7* would be launched first and serve double duty as *Gemini 6*'s target. *Gemini 7* lifted off on December 4, 1965, carrying astronauts Frank Borman and James Lovell. As soon as it had cleared the launch pad, engineers moved in for a new setup. Eleven days later, after two aborted launches, *Gemini 6* took off in hot pursuit. Boosted into a lower orbit, *Gemini 6* caught up with its twin in less than six hours.

Slowly and painstakingly, Schirra used his thrusters to bring *Gemini 6* within six inches of *Gemini 7*. Lacking the Agena's docking equipment, the two ships could not physically link, but they came nose to nose. Lovell in *7* stared across a few feet of space to Schirra in *6*: "I can see your lips moving," said Lovell. "I'm chewing gum," replied Schirra. For seven hours the foursome cracked jokes and shot pictures of the panorama below them. The two capsules floated in close formation for almost three orbits before *Gemini 6* made a perfect manual reentry and splashdown. Left to its own devices, *Gemini 7* continued to circle the Earth for an unprecedented two weeks. After 206 orbits and 5.7 million miles, Borman and Lovell splashed down on December 18, 1965, setting a new record for human space flight. Though sweaty, unshaven, and a bit stooped, the two were little the worse for wear.

## THE PERILS OF DOCKING

With *Gemini 8,* astronauts at last accomplished a true docking. Neil Armstrong and David Scott linked up to an Agena rocket on March 16, 1966. Their victory celebration was cut short, however, when the coupled vehicles began an uncontrolled spin around each other, probably because one of the Gemini thrusters was firing uncommanded. Armstrong quickly detached his capsule from the Agena rocket, but Gemini continued its own dice roll. "We've got serious problems here," hollered Armstrong. "We're tumbling end over end . . . and we can't turn anything off!"

In a last-ditch effort, a dizzy and almost incapacitated Armstrong activated the reentry system. The ship's somersault slowed to a standstill, but *Gemini 8* was left with only a quarter of its precious maneuvering fuel, giving the crew only one chance to reenter Earth's atmosphere. If they missed, they were doomed to permanent Earth orbit. Observers held their breath as the craft fired its retrorockets over southern China. The timing was perfect; *Gemini 8* splashed down a mile from target.

Determined to try the maneuver again, NASA launched *Gemini 9,* only to have technical glitches prevent the rendezvous. Although astronaut Eugene Cernan managed a record two-hour, eight-minute EVA, he did not get to try out the newly developed astronaut maneuvering unit (AMU), a backpack that was nicknamed the "flying armchair." Once outside, as he tried to make his way to the rear of the capsule, where the AMU was located, he had to fight so hard to keep his weightless body from simply drifting away every time he touched the spacecraft that his spacesuit could not deal with the body heat he was generating. By the time he sat himself down in the AMU, his helmet had fogged up and blinded him. The Gemini program clearly had a long way to go before its astronauts would master the tricky techniques required for Moon missions.

Finally, *Gemini 10* solved the rendezvous problem, successfully hooking up with *Agena 10* on July 18, 1966. Astronauts aboard the spacecraft then tracked down the Agena abandoned by *Gemini 8,* planning to use the booster as a power pack to carry them into a new orbit. After docking, Michael Collins punched in the firing command codes and was dismayed to see a thin trail of snowball-like objects dribble out of the Agena's stern. As Collins recalls, he was about to report a malfunction when "wham!—the whole sky turns orange-white and we are plastered against our shoulder straps." In fourteen seconds, the Agena booster had lifted the twin vehicle into a record-breaking 475-mile-high orbit.

*Gemini 11,* lifting off almost two months after *Gemini 10,* also linked with an Agena and repeated *Gemini 10'*s boosting maneuver, doubling the height of the previous orbital record to 850 miles. The crew was so high up that, for the first time ever, they could see the planet in its entirety. "The world is round!" shouted Pete Conrad.

Weary mission planners knew that major obstacles to true space flight had been surmounted. Gemini crews had shown that they could survive for weeks. Spacecraft could be launched and steered with pinpoint precision; even after crossing tens of thousands of miles of featureless vacuum, two craft could meet and dock. Yet one barrier remained: the accursed EVA. No lunar voyage could succeed if the crew could not leave the landing craft.

Jim Lovell and Edwin "Buzz" Aldrin, the crew of *Gemini 12,* took off in November of 1966 determined to solve this final problem. In fact, the two men walked to the launch pad with signs reading "THE END" taped to their backs. The mission began smoothly, but the astronauts had trouble docking with *Agena 12* when their radar faltered. Demonstrating that old-fashioned methods are sometimes the best, Lovell maneuvered the Gemini capsule in by hand, while Aldrin navigated with a sextant. Once the vehicles were securely attached, Aldrin attempted to break the EVA jinx once and for all. Wearing a harness and special gloves that adhered to patches on the outside of the spacecraft, Aldrin spent a total of 128 minutes doing chores like cleaning Lovell's window. By the end of the mission, he had logged a total of five and a half hours outside the capsule without tiring, proving at last that extended EVAs were not impossible.

With a near-perfect splashdown—about two and a half miles from its target—*Gemini 12* was a grand finish to the program. Between March 23, 1965, and November 15, 1966, Gemini had logged ten piloted missions and 1,940 person-hours in space at a cost approaching $1.3 billion. In those twenty months, the program had closed the gap in a race the Soviets had led for a decade. More important, the two countries together had proved that human beings had the technological capability—and the physical and psychological endurance—to survive the long and perilous trek to the Moon.

# CHARTING A COURSE
# THROUGH THE VOID

**B**uck Rogers, the space-traveling hero of science-fiction epics, made space flight seem easy. Whenever he wanted to go to the Moon, he had only to climb into his ship, grab the steering wheel, and blast off, with his rocket motor blazing until he got there. But a real lunar expedition is a highly complex navigational exercise. In contrast to a drive on Earth, where destinations stay put, the trip involves targets that move in three dimensions and at high speed: The Moon orbits the Earth at roughly 2,300 miles per hour.

The aiming problem is complicated by the impracticality of steering a straight line through the shifting crosscurrents of gravity in space. The combined pull of the Earth and the Moon varies as a spacecraft's position between them changes. Thus a real trajectory must be a series of long curves and loops, all precisely worked out according to Newton's laws of motion. Furthermore, since every precious ounce of fuel must be brought along from Earth, engines are used only when absolutely necessary, and most of the journey is spent coasting. Finally, a jaunt into space, like any other technological undertaking, is subject to mistakes and glitches of all kinds. The length of the trip means that even a minor error in launch velocity or direction at the beginning may lead to gross inaccuracy by the end. Because of the ever-present chance of a major malfunction, any plan for a trip should include opportunities to pause, check things out, and make corrections.

As yet astronauts need not make the journey unaided. Apollo spacecraft, which provide the basis for the lunar voyage shown on the following pages, completed the trip with the help of a worldwide network of telemetry and radar tracking stations, sophisticated computers and communications gear, and the efforts of thousands of people back on Earth.

# Way Points on a Journey to the Moon

Considerations of safety and economy dictate that a voyage to the Moon be performed in stages, since a direct trip leaves no margin for error: Unless everything works perfectly, the spacecraft could smash into the Moon at thousands of miles per hour.

The first stage takes the craft to a parking orbit, about a hundred miles above the surface of the Earth, where astronauts and ground-based controllers can inspect their equipment before the mission proceeds any further. If everything checks out, they fire the ship's rocket engine for the so-called translunar injection, boosting their speed to about 25,000 miles per hour—nearly seven miles per second. After the engine cuts off, the spacecraft spends three days coasting to the Moon under the influence of gravity. This phase of the journey provides plenty of time to verify the course and to make corrections. The path is planned as a free-return trajectory *(box, opposite)*, which means that in the absence of any additional rocket firings, the Moon's gravity would whip the craft around and back to Earth for reentry.

As they pass behind the Moon, the astronauts again fire the engine to slow down and drop into lunar orbit. If all goes well, they can survey the landing site and check out their lunar landing module before descending to the surface.

**1** The spacecraft fires its engines to increase speed and break out of the parking orbit around Earth. Ignition takes place on the side of Earth opposite the position where the Moon will be when the spacecraft arrives. The ship's trajectory to the Moon lies in the same imaginary plane as its parking orbit around Earth.

Trans-Earth Trajectory

Earth

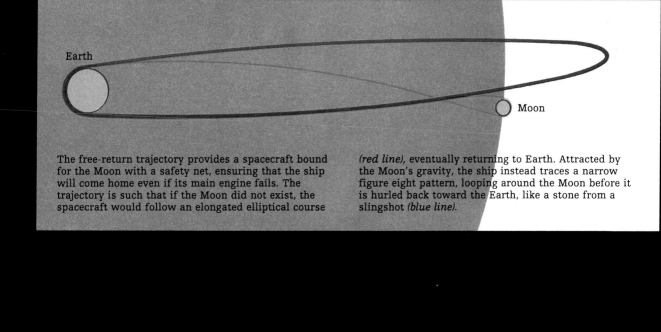

The free-return trajectory provides a spacecraft bound for the Moon with a safety net, ensuring that the ship will come home even if its main engine fails. The trajectory is such that if the Moon did not exist, the spacecraft would follow an elongated elliptical course *(red line)*, eventually returning to Earth. Attracted by the Moon's gravity, the ship instead traces a narrow figure eight pattern, looping around the Moon before it is hurled back toward the Earth, like a stone from a slingshot *(blue line)*.

**3** While astronauts conduct their activities on the lunar surface, the Moon moves along its elliptical path around the Earth. Upon return to the main vehicle, they fire its engines to speed up and leave the lunar neighborhood. The trans-Earth trajectory lies in the same plane as the orbit around the Moon.

**2** Since the spacecraft's lunar orbit may lie in a different plane than that of the translunar trajectory, the rocket burn on reaching the Moon is used to change both speed and direction. As the ship swings behind the Moon under the pull of lunar gravity, the pilots turn it around so that the rocket engine faces forward along the direction of motion. When they cross the plane of their planned orbit around the Moon, the pilots fire the engine, slowing to orbital velocity and changing planes at the same time.

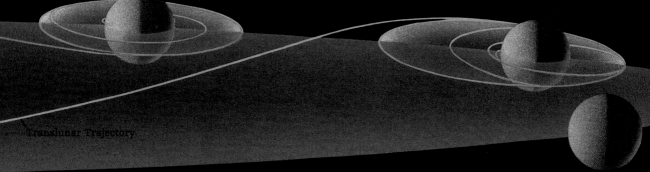

Translunar Trajectory

**Moon Position at
Translunar Injection**

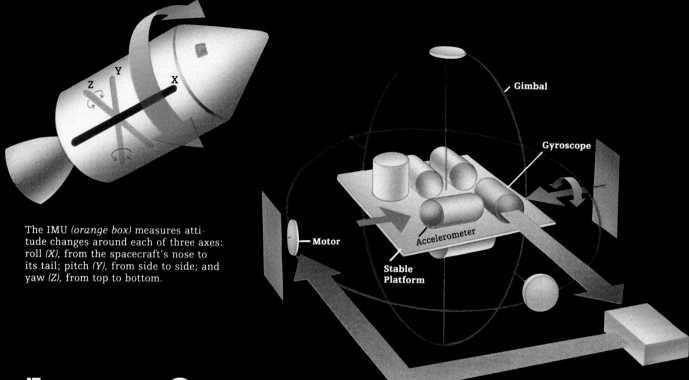

The IMU *(orange box)* measures attitude changes around each of three axes: roll *(X)*, from the spacecraft's nose to its tail; pitch *(Y)*, from side to side; and yaw *(Z)*, from top to bottom.

# FIGURING OUT WHICH END IS UP

Except for brief bursts of rocket power to alter the trajectory, a trip to the Moon is mostly a matter of coasting through space at very high speed. The occupants of an unpowered ship are weightless and therefore cannot tell which way is up or down: Every direction seems the same. Nonetheless, they and their earthbound helpers must keep track of the spacecraft's attitude, its physical orientation in space.

In the Apollo spacecraft, attitude is measured by an instrument called the inertial measurement unit (IMU). At the heart of the IMU illustrated here are three gyroscopes, which operate on the principle that a wheel spinning rapidly around its axis will resist any force that tries to tilt the axis. The gyroscopes reside in cylindrical canisters that are placed at right angles to one another and attached to a so-called stable platform. The platform, in turn, is suspended inside a set of pivots, or gimbals, that allow it to stay motionless while the spacecraft twists around it. The ship's attitude is determined by reference to the orientation of the stable platform.

In this way, the IMU keeps track of the velocity effects of rocket burns and atmospheric effects during reentry. Accompanying each gyroscope is an instrument called an accelerometer that detects any changes in speed along each of the platform axes *(above)*.

When the attitude of the ship starts to change, indicated here by the green double-headed arrow *(above)*, the change in orientation of each gyroscope wheel triggers a signal *(pink arrow)* to an amplifier and then to a corresponding electric motor. The motor, in turn, moves a set of gimbals just enough to restore the wheel to its original orientation. Adjustment of the gimbals keeps the IMU's stable platform motionless.

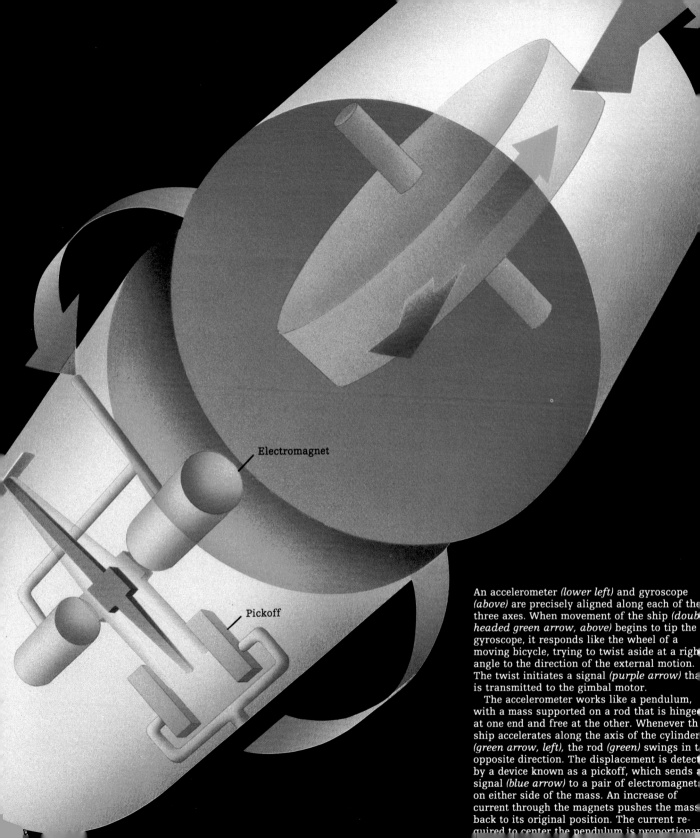

Electromagnet

Pickoff

An accelerometer *(lower left)* and gyroscope *(above)* are precisely aligned along each of the three axes. When movement of the ship *(double-headed green arrow, above)* begins to tip the gyroscope, it responds like the wheel of a moving bicycle, trying to twist aside at a right angle to the direction of the external motion. The twist initiates a signal *(purple arrow)* that is transmitted to the gimbal motor.

The accelerometer works like a pendulum, with a mass supported on a rod that is hinged at one end and free at the other. Whenever the ship accelerates along the axis of the cylinder *(green arrow, left),* the rod *(green)* swings in the opposite direction. The displacement is detected by a device known as a pickoff, which sends a signal *(blue arrow)* to a pair of electromagnets on either side of the mass. An increase of current through the magnets pushes the mass back to its original position. The current required to center the pendulum is proportional

# HIGH-TECH CELESTIAL NAVIGATION

Because gyroscopes slowly drift out of alignment, an unattended IMU could become inaccurate and fatally misleading. To reset the system, astronauts use star sightings, combining the latest technology with methods employed by generations of earthly seafarers. The ship's position and velocity in space are ordinarily measured by ground-based tracking radar. Usually this system gives a fast, precise reading, but in the event of a communications loss, astronauts can fall back on celestial navigation.

This technique relies on the fact that because stars are so distant they seem fixed in place, even on a trip to the Moon. Navigators treat the stars as luminous points on the inside of a so-called celestial sphere. Astronauts can confirm the attitude of their spacecraft in relation to this unchanging backdrop. By combining star sightings with measurements of closer celestial landmarks, such as the horizon of the Earth or a crater on the Moon, they can determine their actual position in space. Exact measurement and timing are critical to these calculations, since Earth, Moon, and spacecraft are always moving. Therefore, the shipboard sextant is linked to the navigational computer, which uses internal star catalogs and its extremely accurate clock to transform the data into a positional fix. Successive fixes can be plotted to show the course the ship has followed and to predict its trajectory.

The celestial sphere—shown here greatly out of scale—is the imaginary surface on which the fixed stars are considered to be located. The sphere's equator, north pole, and south pole are extensions of the same features on Earth. To ascertain the alignment of the IMU and subsequently the attitude of their spacecraft, astronauts must determine the exact angular direction to a pair of stars on the sphere. Finding their position requires that they measure the angle between a star and reference points on the Moon or the Earth. Their on-board computer calculates the one position in space that is consistent with those measurements at the time they were made.

**Celestial Equator**

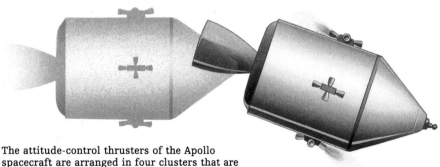

The attitude-control thrusters of the Apollo spacecraft are arranged in four clusters that are spaced at ninety-degree intervals around the circumference of the ship. To turn the nose down, as shown here, the astronauts fire the back thruster on the top of the craft along with the forward thruster on the bottom. Thrusters are always used in opposing pairs for attitude adjustment: If the top one was used alone, for example, it would push the spacecraft forward in addition to turning it.

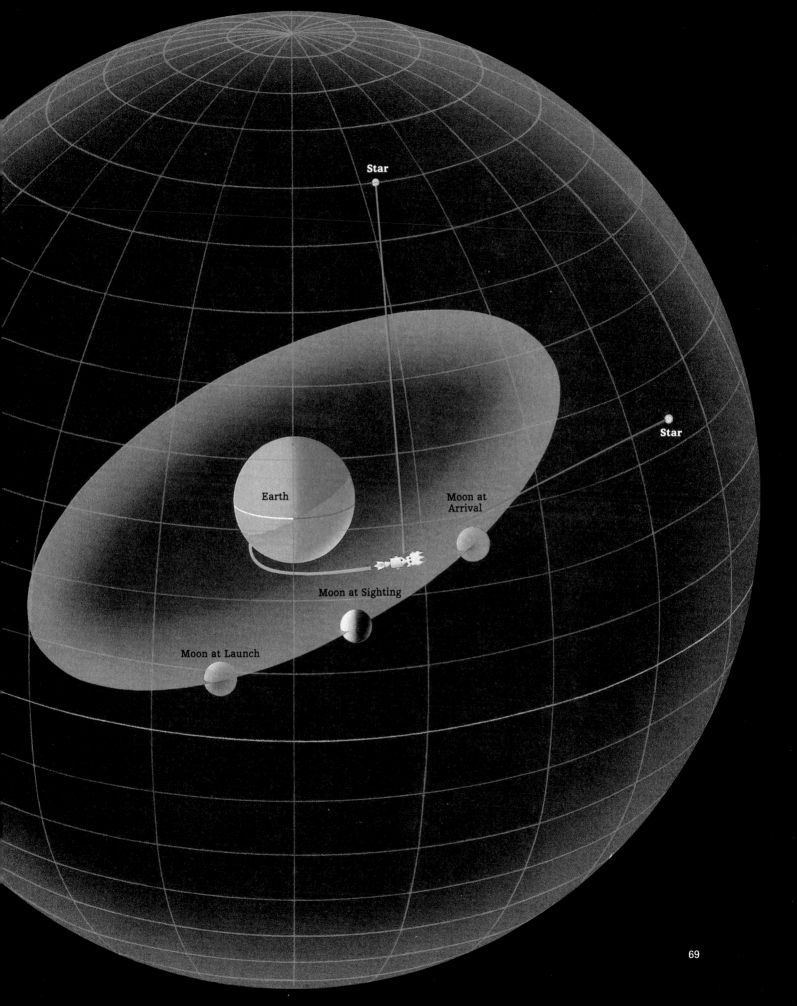

Star

Star

Earth

Moon at
Arrival

Moon at Sighting

Moon at Launch

# GETTING DOWN TO THE LUNAR SURFACE

Before astronauts can land on the Moon, their spacecraft must achieve a lunar orbit. As the inbound ship rounds the far side of the Moon, its pilots slow it by turning it around and firing the main rocket engine. The Moon's gravity then bends their path into an el-

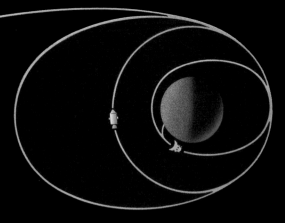

The lander begins its descent in a roughly horizontal position, which was its orientation when it separated from the command module. Starting at an altitude of about ten miles, it will fire its big descent engine continuously until just before touchdown.

At 40,000 feet the pilots begin to monitor their four-beam radar. One beam *(green)* measures the distance to the surface along the diagonal descent path. The other three *(yellow)*, operating at a different frequency, are used to sense the rate of approach to the surface.

liptical orbit, with its high point, or apolune, about 200 miles above the surface on the near side, facing Earth, and its low point, or perilune, 70 miles above the surface on the far side.

This initial orbit provides another breathing space, allowing the astronauts to check the trajectory and verify that everything is going according to plan. The next downward step comes two orbits later: A second engine burn lowers the orbit to ten miles on the near side, while it remains seventy miles high on the far side. After another ten orbits for rest and still more checkout, the landing team enters the lunar lander, separates from the command module, and begins the final descent. The command module moves into a circular orbit at seventy miles. When the lander, still in the ten-by-seventy-mile elliptical orbit, reaches the ten-mile perilune, the pilots fire its engines to reduce their orbital velocity and close in on their target.

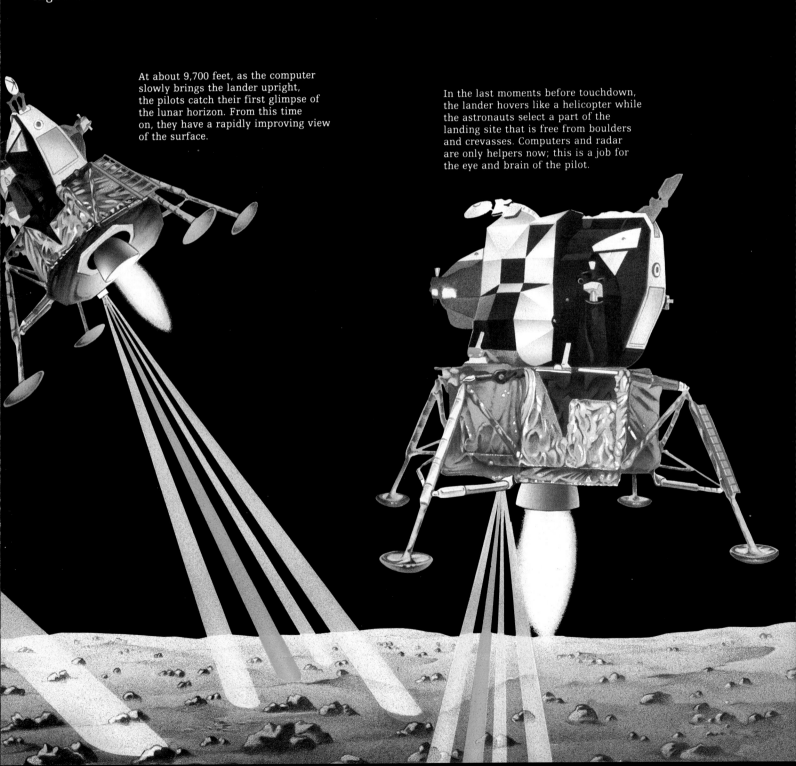

At about 9,700 feet, as the computer slowly brings the lander upright, the pilots catch their first glimpse of the lunar horizon. From this time on, they have a rapidly improving view of the surface.

In the last moments before touchdown, the lander hovers like a helicopter while the astronauts select a part of the landing site that is free from boulders and crevasses. Computers and radar are only helpers now; this is a job for the eye and brain of the pilot.

# Blastoff, Rendezvous, and the Trip Home

The homeward journey from the Moon's surface begins with one of the trickiest phases of the whole mission—the return to lunar orbit and rendezvous with the command module. Like most other procedures, it is done slowly, following a step-by-step scenario, for safety's sake.

The first leg of the trip begins with the blastoff from the lunar surface. The lower half of the lander functions as a launch pad and will stay behind. The upper half, which has a self-contained rocket engine, serves as the ascent vehicle, ultimately carrying the astronauts to an altitude of about seventy miles.

During the ascent, the pilots begin using their rendezvous radar to track the orbiting command module. After the usual position and equipment checks, the lander pilots fire their engine again to propel the craft to the command module's altitude. The docking phase begins when the two craft are in the same orbit, with

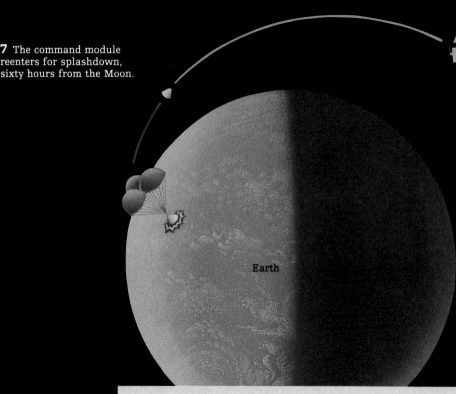

**6** Midcourse corrections refine the aim for a precise Earth landing.

**7** The command module reenters for splashdown, sixty hours from the Moon.

Earth

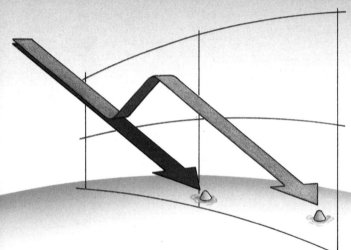

## Precision Skipping

Returning lunar spacecraft can reenter the atmosphere in two ways. One *(red arrow)* is to plow straight in at a diagonal, letting air friction against the heat shield slow the capsule down until it can safely use its parachutes. The other *(pink arrow)* is to come in at a slightly shallower angle so that the spacecraft skips once across the top of the atmosphere. This maneuver calls for very precise guidance, for if the reentry path is even a little too shallow, the capsule could bounce away from the atmosphere entirely *(page 41)*. Skipping can be useful in an emergency: If a last-minute typhoon makes the landing site too dangerous, a well-planned skip will take the spacecraft about a thousand miles downrange to safety.

a fixed distance between them. The lander closes in on the command module, maneuvering with gentle thrusts from its attitude-control jets. In the event of an equipment failure in the lander, the pilot of the command module is prepared to descend and complete the operation.

The return to Earth is almost a mirror image of the outward journey. After the landing team transfers to the big ship, the astronauts jettison the lunar vehicle and fire the command module's propulsion engine for the trans-Earth injection. They spend the next two and a half days coasting, making occasional star sightings and any necessary midcourse corrections. As they approach the Earth's atmosphere, the astronauts jettison their propulsion and support equipment and turn the capsule's blunt heat shield forward to absorb the heat of reentry. A midocean splashdown marks the completion of the journey.

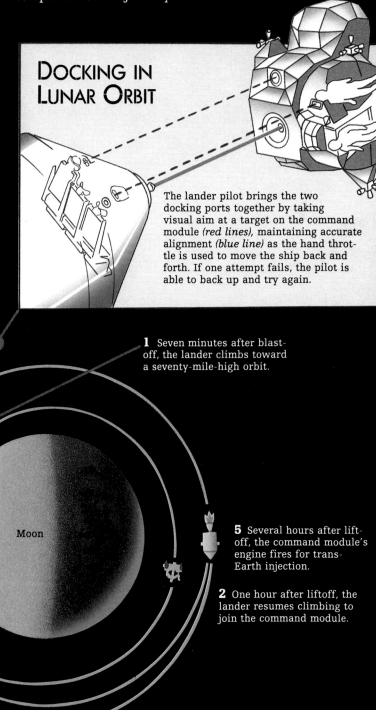

## DOCKING IN LUNAR ORBIT

The lander pilot brings the two docking ports together by taking visual aim at a target on the command module *(red lines)*, maintaining accurate alignment *(blue line)* as the hand throttle is used to move the ship back and forth. If one attempt fails, the pilot is able to back up and try again.

**1** Seven minutes after blast-off, the lander climbs toward a seventy-mile-high orbit.

**3** The lander and command module rendezvous ninety minutes after lunar liftoff.

Moon

**5** Several hours after lift-off, the command module's engine fires for trans-Earth injection.

**2** One hour after liftoff, the lander resumes climbing to join the command module.

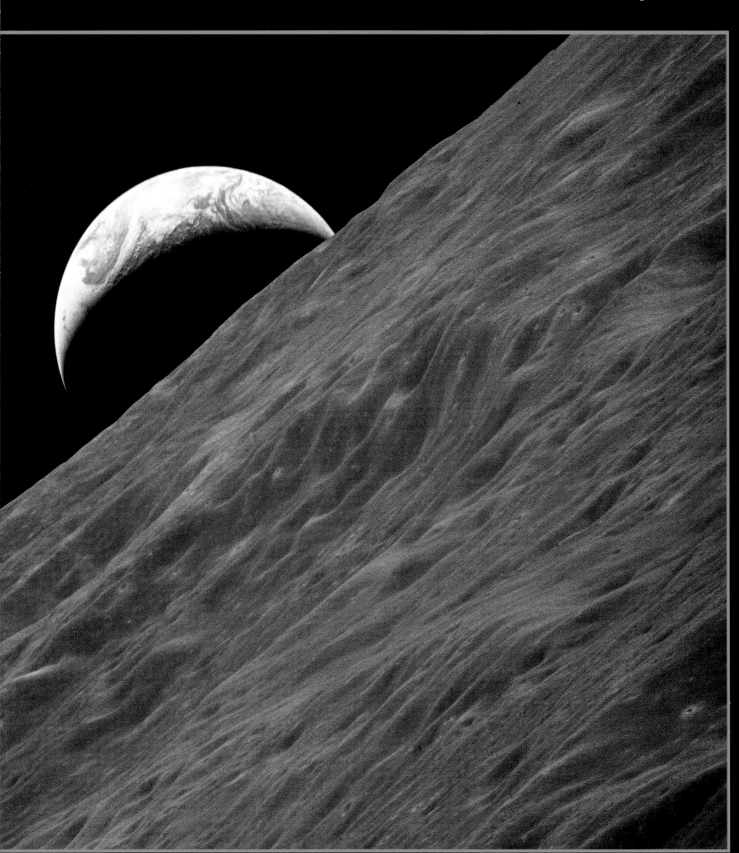

A crescent Earth rises over the dusty, cratered shoulder of the Moon—a view seen by no human eyes since December 17, 1972, when the last Apollo astronauts snapped this shot of home just before leaving lunar orbit.

arly in the morning of Christmas Eve, 1968, a snub-nosed craft bearing the insignia of the United States swung around the back side of the Moon. For the next half-hour, the ship would be out of touch with Earth, its radio transmissions blocked by the Moon's bulk. During this transit—the first passage of a piloted spacecraft behind another body in the Solar System—a crucial maneuver would take place. The three-member crew had to fire the vehicle's main engine for about four minutes to enter into lunar orbit. If the engine fired too long, the ship could crash into the Moon's surface. If it did not fire long enough, craft and crew might go into permanent lunar orbit, beyond hope of rescue.

The flight was a test of the most basic rules of space travel. Until now, no craft had carried humans out of Earth orbit, a necessary precursor to any interplanetary voyage, and none had ever put humans under the primary gravitational influence of another celestial body. If all went well, the mission would also be the first to bring humans back from beyond Earth orbit, reentering the atmosphere at a speed of nearly 25,000 miles per hour—7,000 miles per hour faster than the reentry speed of the Gemini capsules.

For thirty-four nerve-racking minutes, the world waited. Inside the space capsule were two veteran spacefarers and one rookie: Frank Borman, a steely-eyed forty-year-old fighter pilot who had been the commander three years earlier on *Gemini 7;* his Gemini partner Jim Lovell, also forty, who had gone on to command *Gemini 12;* and thirty-five-year-old William Anders, a nuclear engineer making his maiden trip into space. This flight, designated *Apollo 8,* marked only the second time an Apollo craft had successfully lifted off with a crew: Suspense was thus high—and the stakes were huge. The Soviets had already sent a spacecraft containing a "crew" of flies, mealworms, turtles, and other animals to the Moon and back, and they were clearly gearing up for a lunar flight that would carry cosmonauts.

At the instant the ship was due to reappear from behind the Moon, mission control began calling: *"Apollo 8,* Houston, . . . *Apollo 8,* Houston, over." For a few seconds, the listeners in Houston heard nothing but the faint crackle of space. Then Jim Lovell's voice broke the silence: "Go ahead, Houston, *Apollo 8."* The burn had been perfect.

That night, late on Christmas Eve in the United States and early Christmas morning in Europe, *Apollo 8* made a television broadcast from lunar orbit.

"For all the people back on Earth," said Anders, "the crew of *Apollo 8* has a message that we would like to send to you." Anders then began to read: " 'In the beginning God created the heaven and the earth. . . .' " After the first three verses of the Book of Genesis, Lovell took up the reading: " 'And God called the light Day, and the darkness He called Night. . . .' " Finally Borman's turn came: " 'And God said, Let the waters under the heaven be gathered together unto one place, and let the dry land appear: and it was so. And God called the dry land Earth, and the gathering together of the waters he called Seas: and God saw that it was good.' And from the crew of *Apollo 8*," he concluded, "we close with good night, good luck, a Merry Christmas, and God bless all of you—all of you on the good Earth."

## A CHALLENGE WELL MET

*Apollo 8*'s accomplishment and Christmas broadcast moved the whole world. But only NASA was in a position to know just how great the accomplishment was. Nearly eight years earlier, when President Kennedy stepped up the timetable for landing an astronaut on the Moon, NASA's development program had gone into overdrive. As it happened, the rocket power that would be needed was already on the drawing board. Since 1959, Wernher von Braun and his team in Huntsville, Alabama, had been working on an enormous new booster that would eventually become the Saturn V. Towering 364 feet into the air and weighing 18 million pounds when fully loaded, the Saturn V required extraordinary fabrication and launch procedures. Inside a Cape Canaveral structure known as the Vehicle Assembly Building (itself 526 feet high), the components were put together on a mobile platform that would double as the launch pad. The platform, powered by six huge diesel engines, carried the rocket three and a half miles over the Florida countryside to the launch site, moving at a maximum speed of about one mile per hour.

With a thrust of 7.5 million pounds at launch, the Saturn V would prove a solid workhorse, establishing a safety record unmatched by any other family of rockets in the United States or the Soviet Union: In the eleven years the booster launched American spacecraft into orbit, not a single Saturn V ever malfunctioned in any way.

As von Braun's team was working on the rocket itself, North American Aviation, which later became North American Rockwell and then Rockwell International, was building the major part of the rocket's payload: Apollo's command and service modules, known jointly as the CSM. The command module, which would be home for three astronauts during most of their flight, was eleven feet high and thirteen feet across, giving it four times the interior space of Gemini. With more than two million parts, including 566 cockpit switches and 71 lights, it was one of the most complex devices ever built.

Directly beneath the command module on the "stack," as the full Apollo-Saturn complex was known, was the service module, a separate cylindrical section twenty-four feet long and thirteen feet across. The crew could not enter this part of the craft, but it was central to their flight, containing the oxygen tanks and fuel cells that provided them with electricity, oxygen, and water. The service module was equipped with the large rocket engine used to insert the craft into lunar orbit and to project it out of that orbit for the return trip. Sixteen small thruster rockets were designed to maintain the craft's proper orientation, or attitude, in space and to supply the small amounts of thrust needed for midcourse corrections *(pages 63-73)*.

The schedule for building the various Apollo craft placed great burdens on the men and women involved. To meet the president's end-of-the-decade deadline for a lunar landing, NASA had to start testing Apollo in space a bare three months after the last Gemini mission splashed down. With design and construction forced to get under way before the final configuration of the entire spacecraft could be determined, Apollo became a creature of evolution, changing almost weekly as new requirements and limitations arose. The project was rife with technical problems in need of solution, ranging from the design of a fuel gauge that would work in zero gravity to finding ways of preventing the gigantic Saturn rockets from shaking themselves to pieces.

The first piloted Apollo flight, a tryout of the spacecraft in Earth orbit, was originally scheduled for December 1966, but delays in the testing of the command and service modules caused a lengthy postponement. Finally, on January 27, 1967, the CSM was perched atop the Saturn V rocket for launch-pad trials. If all went well, liftoff would take place about three weeks later.

In the weeks leading up to launch, the crew scheduled to fly *Apollo 1*—Roger Chaffee, Gus Grissom, and Edward White—had grown increasingly frustrated with all the glitches. At one point, Grissom had hung a giant lemon on the ship to express his discontent. Even when the three astronauts were sealed in the cockpit and going through a preflight checkout, the troubles persisted. Communications breakdowns were so exasperating that Grissom finally barked, "How the hell can we get to the Moon if we can't even talk between two buildings?"

About five and a half hours into the test, Roger Chaffee reported over the radio in an almost casual tone, "Fire, I smell fire." A surge of movement in the cabin followed, then Chaffee cried, "We've got a bad fire. We're burning up here." Then nothing was heard from the cabin but frantic pounding on the walls and shouting that no one on the launch pad could make out. Sixteen seconds after the first word from Chaffee, a section of the command module burst in a violent explosion of smoke and flame. By the time anyone could reach the crew, the three were dead of asphyxiation.

## AFTERMATH OF TRAGEDY

Only a few weeks earlier, an eerily prophetic Gus Grissom had commented on the risks of space flight in a conversation with news reporters. "We hope that if anything happens to us it will not delay the program," he said. But delay was inevitable. A board of inquiry conducted an exhaustive review of the accident, producing a 3,000-page report that labeled the workmanship in the

The engineers who designed and built the craft that would take the first humans to the surface of the Moon enjoyed an unprecedented advantage. Because the lunar module, or LM, would never fly anywhere but in the vacuum of space, designers could ignore the usual rules of aerodynamics; odd bulges or uneven surfaces would not affect the craft's flight. The resulting lumpy bundle, poised over out-turned legs like a huge metallic insect, deserved its nickname: the bug.

Constraints came from the realities of the launch pad. Since every pound of craft required three pounds of propellant, the ascent stage could weigh in at no more than about five tons. Among other slimming measures, engineers used acids to mill the walls of the LM to the thickness of a credit card, replaced ordinary circuits with narrow-gauge wiring, and eliminated seats for the two-person crew, who would stand during the brief trips between Moon and command module. The weight-loss program had one rather frightening consequence, however. Though durable enough for low lunar gravity, the LM was too fragile to withstand a dry run on Earth. The first full test of this ungainly pioneer would occur when the crew from *Apollo 11* tried to land on the Moon and take off again.

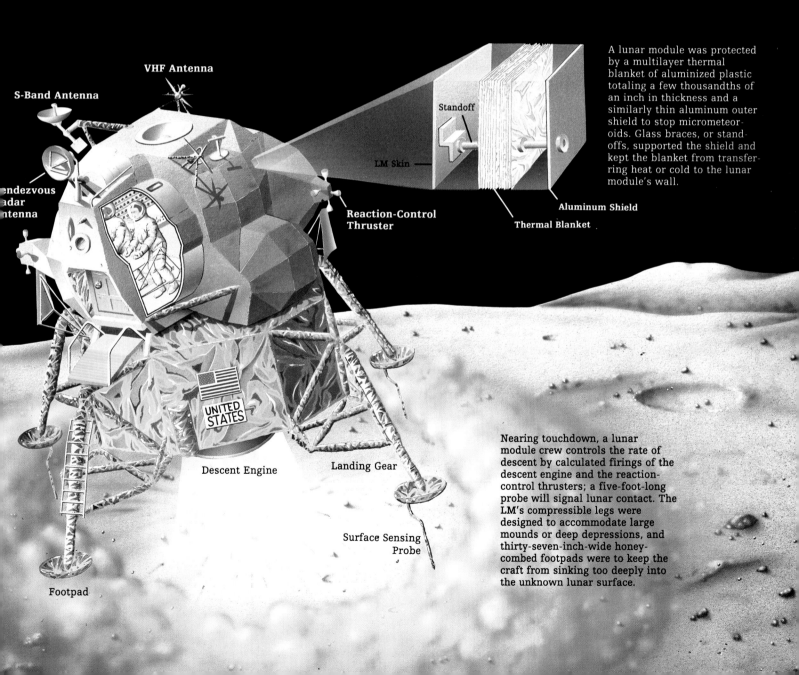

**VHF Antenna**

**S-Band Antenna**

Rendezvous Radar Antenna

**Standoff**

LM Skin

A lunar module was protected by a multilayer thermal blanket of aluminized plastic totaling a few thousandths of an inch in thickness and a similarly thin aluminum outer shield to stop micrometeoroids. Glass braces, or standoffs, supported the shield and kept the blanket from transferring heat or cold to the lunar module's wall.

**Aluminum Shield**

**Thermal Blanket**

**Reaction-Control Thruster**

UNITED STATES

**Descent Engine**       **Landing Gear**

**Surface Sensing Probe**

**Footpad**

Nearing touchdown, a lunar module crew controls the rate of descent by calculated firings of the descent engine and the reaction-control thrusters; a five-foot-long probe will signal lunar contact. The LM's compressible legs were designed to accommodate large mounds or deep depressions, and thirty-seven-inch-wide honeycombed footpads were to keep the craft from sinking too deeply into the unknown lunar surface.

# A Lunar Home and Communications Hub

For the twenty-one to seventy-five hours of a lunar landing mission, the LM was both a refuge from the inhospitable environment and a vital communications center, linking the LM crew, their colleague orbiting in the command module, and ground stations on Earth. During later missions, the system also incorporated transmissions to and from the wheeled lunar rover.

Voice communications took place on S-band, at frequencies from 2,100 to 2,300 megahertz, or on VHF (very high frequency), at 250 to 300 megahertz. S-band carried all dialogue with Earth. Live television coverage of the Moon landings came via S-band, as did biomedical and engineering data that let mission control monitor the life-support and flight systems of both lunar craft. Since the LM's antenna was small, S-band transmissions were also relayed to and from Earth through a much larger antenna that could be erected on the lunar surface. VHF was used for short-range communications among the lunar module, astronauts operating outside the LM, and the command module; it also conveyed biomedical data from the crew's spacesuits during extravehicular activities.

The normal three-way information exchange was easiest when the command module and Earth were both above the Moon's horizon *(right)*. Because landings were set on the side of the Moon that always faces Earth, the LM crew was never out of direct contact with home. The orbiting command module, however, was out of touch with both the landing party and Earth for half of each two-hour orbit.

When the command module was above the horizon, the LM crew talked with it by VHF; otherwise, Earth relayed messages on S-band.

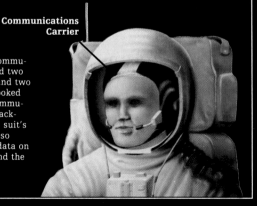

**Communications Carrier**

An astronaut's caplike communications carrier included two independent earphones and two microphones. The cap hooked up with the spacesuit communicator (SSC), part of a backpack device linked to the suit's VHF antenna. The SSC also transmitted monitoring data on the lunar environment and the astronaut's vital signs.

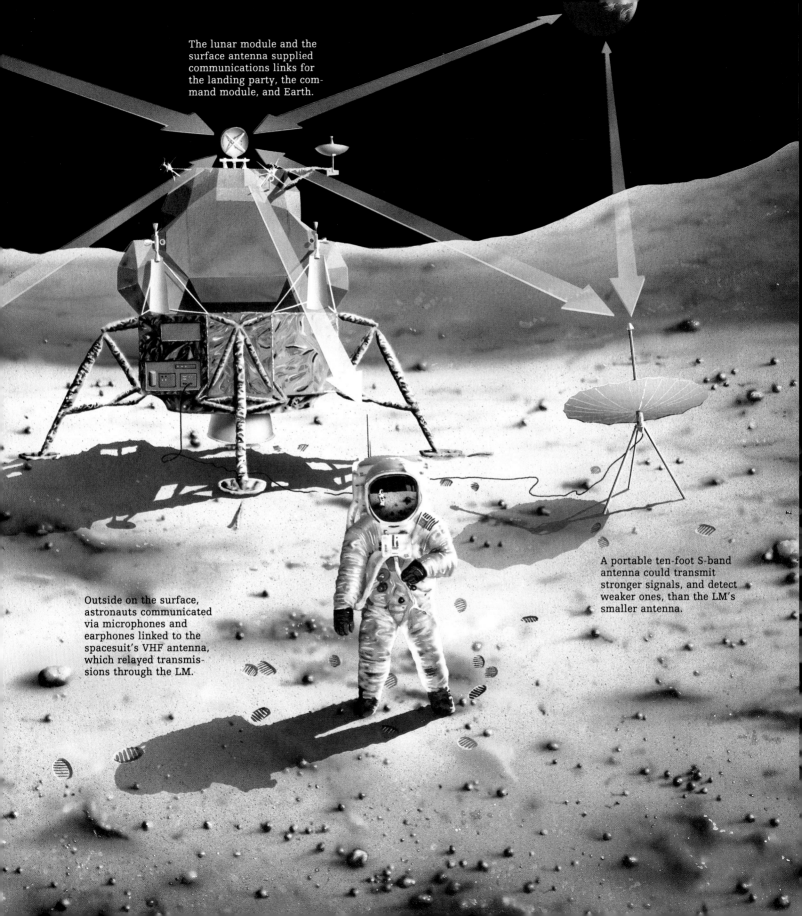

The lunar module and the surface antenna supplied communications links for the landing party, the command module, and Earth.

A portable ten-foot S-band antenna could transmit stronger signals, and detect weaker ones, than the LM's smaller antenna.

Outside on the surface, astronauts communicated via microphones and earphones linked to the spacesuit's VHF antenna, which relayed transmissions through the LM.

When the landing party was ready to return to the command module at the end of a surface mission, the lunar module underwent a dramatic transformation that involved split-second timing. Explosive bolts and a double-bladed guillotine *(box, below)* severed all electrical, life-support, and structural connections between the descent stage, which now became the launch pad, and the 10,582-pound ascent stage, which would carry the crew into lunar orbit for their rendezvous with the command module.

Astronauts and ground controllers back on Earth all viewed this phase of the lunar missions with a certain amount of trepidation. The timing of the separation of the lunar module's two stages was critical; there was no margin for error. Because of the LM's overall weight restrictions, the craft had only one ascent engine. Moreover, all of the remaining fuel was designated for this one blastoff attempt. If something interfered with separation, or if the engine failed, two astronauts would be trapped on the Moon.

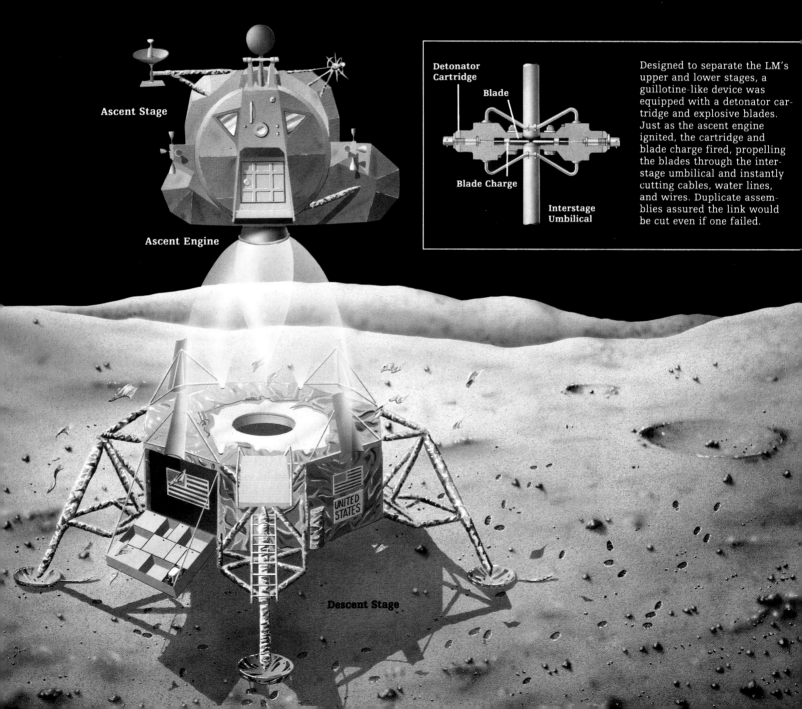

**Ascent Stage**

**Ascent Engine**

**Detonator Cartridge**

**Blade**

**Blade Charge**

**Interstage Umbilical**

Designed to separate the LM's upper and lower stages, a guillotine-like device was equipped with a detonator cartridge and explosive blades. Just as the ascent engine ignited, the cartridge and blade charge fired, propelling the blades through the interstage umbilical and instantly cutting cables, water lines, and wires. Duplicate assemblies assured the link would be cut even if one failed.

**Descent Stage**

UNITED STATES

command module below standard and condemned the test procedures of both North American and NASA. Although no one could pinpoint the exact cause of the fire, investigators concluded that wires in the command module had shorted, generating a spark that set some insulation on fire. In the cabin's pressurized, pure-oxygen atmosphere, the spacecraft material had burned explosively.

NASA's first response to the tragedy and the inquiry board's report was to scrap its compressed launch schedule. Then, over the next year and a half, the agency instituted more than 1,800 alterations in the command module, including the replacement of combustible materials inside the cabin and in the astronauts' spacesuits with flameproof fabrics. In a procedural change, NASA ruled that for all future ground tests and launch countdowns, the cabin's pure-oxygen atmosphere would be a less volatile mixture of 60 percent oxygen, 40 percent nitrogen, changing to pure oxygen only after the flight was under way.

The command module was not the only aspect of the Apollo program to benefit from the more measured pace. The added time in the schedule was especially welcomed by the engineers who were designing the vehicle that would actually land on the Moon.

## ZEROING IN ON A LANDER

The task of inventing an entirely new kind of craft—one that could touch down on the unknown lunar surface and take off again—was not easily accomplished in a few short years. Early in the planning, two possible ways of getting to the Moon had been considered. One, known as direct ascent, was the brute force approach: A spaceship would blast off from the Earth, travel to the Moon, land, and return to Earth. But such an all-purpose ship would need big engines and lots of fuel and would thus require a launch booster much larger than anything that could be completed before the end of the decade. A second option was known as Earth orbital rendezvous. According to this plan, a large spaceship would be lofted into Earth orbit by a Saturn rocket; the necessary extra fuel would be launched separately. The ship would be fueled in orbit, at which point, like the craft for direct ascent, it would head for the Moon, land, and return.

The option that was finally adopted, known as lunar orbital rendezvous, was the brainchild of John Houbolt, an obscure NASA engineer who chaired a small study group on space rendezvous stratagems. Resisted vigorously at first by the Apollo engineering team, Houbolt's plan called for a spaceship to be sent into orbit around the Moon. Next, like a landing boat going ashore from a galleon, part of the craft would separate and descend to the Moon's surface. When the surface mission was completed, the landing vessel would take off from the Moon and rendezvous with the mother ship. The landing craft would then be jettisoned to reduce overall mass—and thus fuel needs—before the mother ship left lunar orbit.

In this way was born the lunar module, or LM, the first true spacecraft.

Because it would fly only in the airless vacuum, the LM did not have to be aerodynamically shaped; in fact, it resembled nothing so much as an ungainly insect *(pages 79-82).*

The delay caused by the *Apollo 1* fire also gave the astronauts time for training in geology and planetary sciences, as well as time to practice in simulators—mock-ups of the command and lunar modules accurate down to the last switch and warning light. Outside the windows of the simulators, computer-controlled arrangements of mirrors and movie projectors displayed what the astronauts would see in space. As the astronauts "piloted" the simulators, their instructors would introduce malfunctions into the equipment, such as an oxygen leak, a power loss, a stuck thruster. The astronauts would quickly have to deal with the problem—or suffer the simulated consequences.

## RETURN TO SPACE

By the latter part of 1968, after several test flights of the Apollo hardware, NASA was again ready for human pilots. Wally Schirra, Walter Cunningham, and Donn Eisele would fly the revamped craft, becoming the first Americans in space in nearly two years.

Except for the fact that all three crew members came down with head colds, the flight of *Apollo 7* was virtually perfect. As intended, the crew took the CSM through its paces in Earth orbit, practiced rendezvousing with the second stage of the booster rocket, and spent nearly eleven days circling the globe before splashdown.

The next three flights were ever more ambitious exercises. After *Apollo 8*'s successful Christmas trip around the Moon, *Apollo 9* drew the assignment of testing the lunar module in Earth orbit. (Neither *Apollo 7* nor *Apollo 8* had carried the LM into space.) On March 3, 1969, the awkward lunar lander rode folded up in the hollow neck of the Saturn V's third stage. Once in orbit, the astronauts performed a tricky maneuver: Disengaging from the third stage, they turned the command module 180 degrees, docked nose to nose with the lunar landing vehicle, and then withdrew it from its shelter. The third stage then drifted away, leaving the command and lunar modules locked together as one ship. Two of the crew entered the LM while one remained in the command module. The lunar module then separated, changed orbit, and dropped as much as 100 miles behind the mother ship before rendezvousing and redocking. Eleven weeks later, *Apollo 10* performed the same rendezvous maneuvers in lunar orbit, the LM's proper arena. While astronaut John Young remained in the command module, Tom Stafford and Gene Cernan brought the lunar module to within 50,000 feet of the Moon's surface.

# APOLLO'S FRATERNITY

In 1961, when the American space program was barely under way, President John F. Kennedy challenged the National Aeronautics and Space Administration to land a person on the Moon by the end of the decade. With only five months and eleven days remaining in 1969, astronauts Neil Armstrong and Edwin "Buzz" Aldrin set foot upon lunar soil.

Their mission was historic, but it was only one item on the four-year agenda of Project Apollo. In all, twelve U.S. astronauts walked—and drove—on the Moon. Not only did they explore a new land, but they also amassed invaluable scientific data, from collections of lunar rocks to measurements of moonquakes. Pictured at right and on the succeeding pages is Apollo's select fraternity of space pioneers.

## FLIGHT OF THE EAGLE

By mid-1969, as much of the equipment and procedures as could be tested ahead of time had undergone trial and been judged satisfactory. The next mission, *Apollo 11,* would attempt to land on the Moon. The day scheduled for the launch, July 16, 1969, dawned hot and clear. Atop the steaming Saturn V—fully loaded with kerosene, liquid hydrogen, and liquid oxygen—were the three astronauts who would attempt to make history. In the left seat of the command module, *Columbia,* was the ship's commander, Neil Armstrong. A licensed pilot at sixteen, he was considered by his colleagues one of the best test pilots in the world, known for his calm during crises and his rational approach to problems. He had flown seventy-eight combat missions in Korea, once escaping from behind enemy lines when shot down.

In the center was Buzz Aldrin, a collegiate pole-vaulter, Korean War fighter pilot, and the author of a doctoral thesis on orbital rendezvous. Three years earlier, Aldrin had been copilot on *Gemini 12* when the rendezvous radar had failed. With the help of information from the ground, Aldrin had made backup calculations that allowed the mission's scheduled rendezvous with the target Agena rocket to proceed. In early flight plans for *Apollo 11,* Aldrin was the designated candidate for first man on the Moon, but by the time of launch, Armstrong, as commander of the mission, had been selected for the initial excursion. Many astronauts believed Aldrin to be bitterly disappointed.

In the right seat was Michael Collins, a test pilot and the reigning handball champion among the astronauts. He professed to be more interested in "girls, football, and chess" than in planes. Collins was the command module pilot, the one destined to stay behind while his crewmates descended to the Moon.

That the flight would be made by these three astronauts among the several

**Apollo 7, October 1968.** After several preliminary missions, Walter Cunningham, Donn Eisele, and Walter Schirra went up in the first piloted flight of Apollo. Their eleven-day voyage tested the spacecraft's operation.

**Apollo 1, January 1967.** Project Apollo began in disaster when astronauts Gus Grissom, Roger Chaffee, and Edward White *(left to right)* died in a capsule fire, one month before their flight was scheduled for launch.

dozen then in the space program was almost pure happenstance. When the crew assignments were laid out more than two years earlier, no one was sure which Apollo mission would attempt the Moon landing. If any problems with the spacecraft had arisen during development and testing, delays would probably have pushed the lunar landing to the next flight, *Apollo 12.* As one astronaut put it, "We had as much control over our futures as lab mice."

At 9:32 a.m., the giant booster roared to life and rose from the launch pad on a red-orange tongue of flame. Inside the command module, the sensation of motion was almost imperceptible at first, increasing as the Saturn V picked up momentum and the five huge rocket engines steered a curving course into space. Collins later likened the feeling to that of "a nervous novice driving a wide car down a narrow alley." For the first few minutes, the astronauts had no view to speak of: Most of the windows in the command module were covered by the escape tower mounted on top of their capsule. But at an altitude of sixty miles, the tower was jettisoned, revealing a blue sky fading to black.

Within twelve minutes of liftoff, the third stage shut down, and *Apollo 11* was in orbit 110 miles above the Earth. The crew spent the next couple of hours inspecting the craft and making navigational checks; then they reignited the third stage to send them speeding toward the Moon. Their eventual target was the western edge of an area called the Sea of Tranquillity, just a little to the right of the Moon's center as seen from the Earth.

After executing a routine docking of the CSM with the lunar module and then abandoning the third stage, the astronauts put their craft into the

**Apollo 9, March 1969.**
Russell Schweickart, David Scott, and James McDivitt, in shirt-sleeves after their splashdown, first tested the separation and docking of the lunar module.

**Apollo 8, December 1968.**
The first humans to fly over the far side of the Moon, James Lovell, William Anders, and Frank Borman circled the satellite for twenty hours, scouting future landing sites.

standard "barbecue" roll, a maneuver that rotated the ship once every twenty minutes to distribute the Sun's heat evenly over its surface. Their housekeeping chores done for the moment, they were free to watch Earth and its moon rise and set in their windows, the former growing ever smaller, the latter ever larger. By the fourth day of the trip, the Sun-dappled body completely filled the view. Again, mission control had to wait anxiously while *Apollo 11* fired its engine on the far side of the Moon to enter lunar orbit. But when the ship emerged from behind the Moon's rim, Collins reported that the burn had been flawless.

Once in lunar orbit, the astronauts were scheduled to rest for a full nine hours in preparation for the next day's historic activity. But all three had trouble falling asleep. Perhaps the most nervous was the one who would not be descending to the surface, Michael Collins. Convinced that the odds for a successful landing and return of the lunar module were no better than even, he was terrified at the prospect of having to leave his crewmates behind, stranded on the Moon or in lunar orbit. Even if he made it back to Earth, he would always consider himself a marked man.

On their tenth trip around the Moon, Armstrong and Aldrin entered the lunar module, which had been named *Eagle,* and two orbits later they backed away from the command module. They then fired *Eagle*'s descent engine to

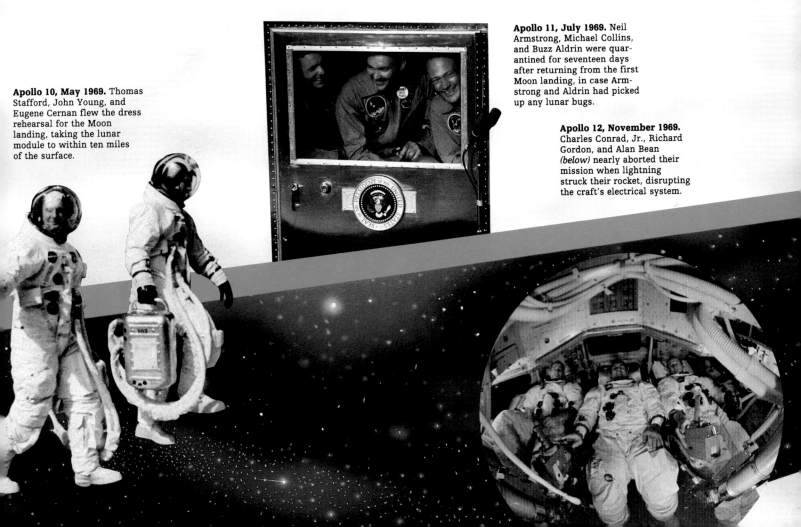

**Apollo 10, May 1969.** Thomas Stafford, John Young, and Eugene Cernan flew the dress rehearsal for the Moon landing, taking the lunar module to within ten miles of the surface.

**Apollo 11, July 1969.** Neil Armstrong, Michael Collins, and Buzz Aldrin were quarantined for seventeen days after returning from the first Moon landing, in case Armstrong and Aldrin had picked up any lunar bugs.

**Apollo 12, November 1969.** Charles Conrad, Jr., Richard Gordon, and Alan Bean *(below)* nearly aborted their mission when lightning struck their rocket, disrupting the craft's electrical system.

drop to a lower orbit. At an altitude of ten miles, or 52,800 feet, they fired the descent engine again. The plan was for the engine to burn continuously for twelve minutes until *Eagle* was safely on the ground.

The first five minutes of the burn went fine. Then an alarm flashed on the computer readout of the command module and on the monitors back in Houston: The LM's computer was overloading. Armstrong and Aldrin did not have time to figure out what was wrong, so the responsibility for deciding whether to abort the mission fell on a twenty-seven-year-old computer specialist in Houston named Steve Bales. Although he later admitted that he was scared to death, Bales concluded that *Eagle*'s computer problem was not a danger to the mission. In a calm, assured voice, he said, "We're go on that alarm," meaning that Armstrong should continue the descent. It was one of the most important decisions made during the entire flight of *Apollo 11,* and for his courage Bales would later join the three astronauts in receiving the Medal of Freedom from President Richard Nixon.

About 200 feet above the surface, Armstrong realized that the computer was going to bring them down in a field of large boulders surrounding a crater. As was planned in any case, he took manual control of the craft and headed toward a smoother site he could see downfield. A light signaled that fuel remained for only about two more minutes of firing time. The veteran pilot pressed on. With less than thirty seconds of fuel remaining, a blue light flashed in the module: The footpads had made contact. "Houston," Armstrong reported, "Tranquillity

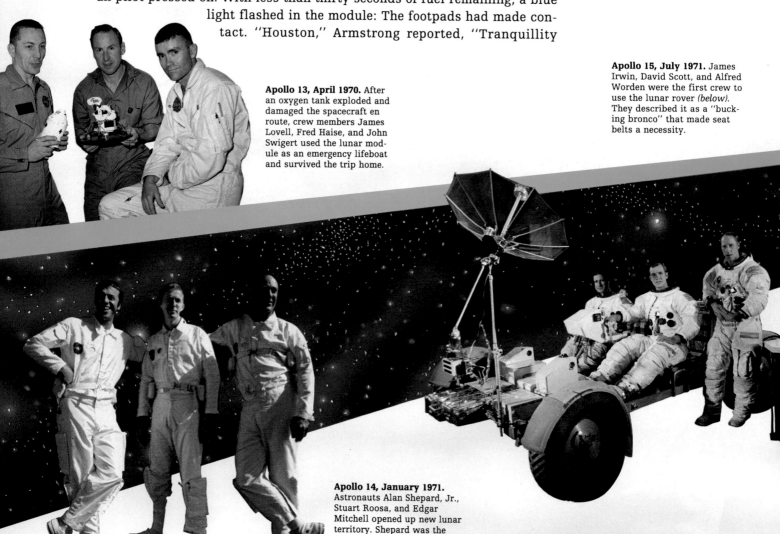

**Apollo 13, April 1970.** After an oxygen tank exploded and damaged the spacecraft en route, crew members James Lovell, Fred Haise, and John Swigert used the lunar module as an emergency lifeboat and survived the trip home.

**Apollo 15, July 1971.** James Irwin, David Scott, and Alfred Worden were the first crew to use the lunar rover *(below).* They described it as a "bucking bronco" that made seat belts a necessity.

**Apollo 14, January 1971.** Astronauts Alan Shepard, Jr., Stuart Roosa, and Edgar Mitchell opened up new lunar territory. Shepard was the only one of the Mercury seven to walk on the Moon.

Base here. The *Eagle* has landed." It was 4:17 p.m., eastern daylight time, on July 20, 1969. The end of the decade was less than six months away.

## ONE SMALL STEP

Although scheduled for some sleep after landing, the astronauts were so keyed up that they requested permission to begin preparations for their excursion onto the surface. Equally keyed up ground controllers in Houston readily agreed. Eager as they were to get outside, the two astronauts spent more than six hours donning their protective suits, checking out their equipment, and depressurizing the lunar module. Finally, Armstrong backed out of the module's hatch. He climbed partway down the nine-step ladder and paused to pull a D-ring that opened a compartment on the side of the ship that deployed a television camera. As hundreds of millions of people on Earth watched, Armstrong took his final step off the ladder onto lunar soil. "That's one small step for man," he said—inadvertently dropping the "a" he had meant to put before "man"—"and one giant leap for mankind."

Armstrong immediately collected a few rocks as a contingency sample, in case they had to leave the Moon in a hurry. He was soon joined by Aldrin, who had been waiting impatiently in the hatch. For the next few minutes, the two astronauts played with the novelty of lunar gravity; they tried out various ways of moving, including hopping and something called the "kangaroo lope." They then unveiled a plaque on the side of the lunar module that read, "Here men from the planet Earth first set foot upon the Moon, July

**Apollo 16, April 1972.**
Charles Duke, John Young, and Thomas Mattingly II road-tested the rover in a kind of lunar grand prix, driving around in circles and skidding to test wheel grip.

**Apollo 17, December 1972.**
In the last Apollo mission, Eugene Cernan, Ronald Evans, and Harrison Schmitt set records for length of stay, distance traveled on the Moon's surface, and number of lunar rocks collected.

1969 A.D. We came in peace for all mankind." Shortly thereafter, they received a telephone call from President Nixon, who called it "the most historic telephone call ever made."

Their ceremonial duties over, the astronauts turned to scientific chores. Besides gathering soil and rock samples, they deployed a number of instruments that would remain on the Moon after their departure. One was an experiment designed to capture charged subatomic particles streaming out from the Sun. Another device, a laser reflector, could be used to measure the distance between the Earth and Moon to within six inches. In addition, they set up a seismometer, which immediately began recording the shocks of their footfalls and returning the data to Earth. Like any pair of terrestrial tourists, they also took a number of pictures—almost all of Aldrin, as it turned out, since Armstrong usually had the camera. Finally, two and a half hours after leaving the lunar module, they climbed back through the hatch and sealed it.

Once again the schedule called for a period of sleep, but the lunar module was too cold and cramped to be restful. After five hours, Armstrong and Aldrin began getting ready to leave. Assuming they succeeded in doing so, the departure would mark the first time any spacecraft—occupied or not—took off from another world. If the ascent engine failed, the pair faced certain death. But Houbolt's ungainly brainchild performed perfectly. Using the descent stage as a launch pad, the ascent stage rose in a plume of exhaust and headed toward the command module orbiting above.

From the moment the ascent stage lifted off, Michael Collins was glued to the window looking for his crewmates. Armstrong's long descent had made it difficult to calculate exactly where the Eagle had landed, so Collins was not sure where to look. Finally, he saw a blinking light. As he watched, the golden bug, sunlight glinting off its protective shield, climbed up from the crater fields below. Only then did he believe that they were really going to pull it off.

The docking was accomplished by hooking the craft together with three small latches and tightening the bond with a rigid connection. As Aldrin, his face split by a huge grin, came through the tunnel joining the lunar and command modules, Collins resisted the urge to grab him by the temples and plant a big kiss on his forehead, settling instead for a handshake. The men proceeded to transfer forty-six pounds of lunar rocks and soil into the command module. Then the Eagle, the trusty bird that had carried the first of the human species to the Moon, was set free. Three days later, tired, smelly, and homesick, the astronauts plopped into the Pacific thirteen miles from the recovery carrier *Hornet.*

## A LUNAR LABORATORY

The rocks and soil returned from the Moon have been called the most expensive geologic samples ever gathered, but to terrestrial scholars they were worth every penny. For the first time, scientists could study materials from another world—and, as a bonus, perhaps learn more about the origins and evolution of Earth.

The rocks that were brought back from the Sea of Tranquillity proved to be extremely ancient. Radioactive dating indicated they had cooled from molten lava between 3.5 and 3.7 billion years ago, making them older than all but a very few rocks on Earth. Because of geologic processes such as erosion, sedimentation, and volcanism, mineral evidence of Earth's first billion and a half years is exceedingly rare. Planetary scientists hoped that the Moon would turn out to be a storehouse of relics from the earliest eras of the Solar System.

The *Apollo 11* samples were merely an appetizer. With *Apollo 12*, the real scientific payoff would begin. Since the flight of the *Eagle* had demonstrated that human pilots could fly the lunar module with pinpoint accuracy, mission designers established an ambitious goal for *Apollo 12:* to find an American robot probe named *Surveyor 3*, which had set down in the area known as the Ocean of Storms in 1967 as part of a program to prepare for Apollo's landings. Before the Surveyor series and a comparable series by the Soviets, no one had been sure whether the Moon's surface would support a craft from Earth; perhaps the bedrock was covered by a thick layer of dust that would swallow a lander like quicksand. But the television camera on Surveyor had revealed lunar dust to be reassuringly similar to dirt on Earth. Now the valiant probe was sitting lifeless on the Moon inside of what the last images from its cameras had revealed to be a shallow crater. *Apollo 12* set out to join it.

**SURVIVING A DOUBLE STRIKE**

The mission began in alarming fashion. Moments after it was launched into a thick cloud cover over the Cape, *Apollo 12* was struck by lightning—first at 6,000 feet and again at 15,000 feet. As overload sensors automatically disconnected the fuel cell batteries, the spacecraft switched to backup power. The strike had caused four fuel gauge sensors to burn out and had thrown the inertial navigation system off kilter. Luckily, these were transient effects: The crew easily realigned the guidance unit and restarted the fuel cells. Ground control's analysis indicated that the craft was undamaged, so once they were in Earth orbit the astronauts were given the green light to carry on to the Moon.

The landing was carefully designed to follow landmarks to the crater that was Surveyor's resting place. As commander Pete Conrad guided the lunar module toward the ground, he could see out the window the hills and craters he had studied in photographs for a period of many months. After a slightly rough landing, which was the result of cutting off the engine a little early, Conrad emerged from the lunar module. "Boy, you'll never believe it," he reported to mission controllers. "Guess what I see sitting on the side of the crater? The old Surveyor.... It can't be any farther than 600 feet from here. How about that?"

Conrad and his partner Alan Bean walked on the Moon for a total of nearly eight hours, more than three times as long as Armstrong and Aldrin's ex-

# CANDIDATES FOR A JAUNT ON THE MOON

As piloted lunar missions neared in the 1960s, engineers dreamed up a fleet of unlikely-looking vehicles that would let astronauts traverse the Moon's surface faster and farther than they could on foot. The proposed conveyances—some shown here and on the next three pages—would face formidable conditions: 500-degree temperature swings, gravity one-sixth that of Earth, pervasive dust, and a vacuum that evaporates most lubricants. Then, in mid-decade, NASA ruled out anything heavy enough to need its own Saturn V launch rocket. Safety and cost concerns eliminated all but a few of the candidates that met the weight test. NASA chose the winner for its combination of lightness, reliability, and fuel economy: Boeing lunar rovers served the crews of *Apollos 15, 16,* and *17.*

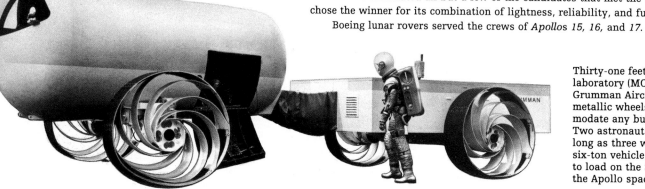

Thirty-one feet long, the mobile lunar laboratory (MOLAB) prototype from Grumman Aircraft rests on five-foot metallic wheels that flatten to accommodate any bumpy terrain and soft s Two astronauts could have spent as long as three weeks at a time in the six-ton vehicle, which proved too hea to load on the Saturn V rocket carryi the Apollo spacecraft.

cursion. On the second of two separate forays from the lunar module, they walked over to Surveyor and cut several pieces from it to take home for study. They also set up the Apollo lunar surface experiments package, or ALSEP, an experimental station that every future Apollo flight would carry. Powered by a nuclear energy device fueled with more than eight pounds of plutonium-238, this particular ALSEP included a seismometer, a magnetometer, and various kinds of charged particle detectors.

Readings from the seismometer deployed by *Apollo 12,* in addition to data from the one left by *Apollo 11,* would allow geologists to measure the intensity and characteristics of "moonquakes," thereby, they hoped, giving them insights into one of the most controversial lunar issues: Is the Moon hot like Earth, with a molten interior, or is it cold and relatively solid all the way through? Scientists also used the *Apollo 12* seismometer to conduct an unusual experiment. As *Apollo 12* was on its way back to Earth, the abandoned lunar module fired its engine one last time and crashed into the Moon's surface, creating an artificial moonquake. To the geologists' amazement, the shock waves from the impact reverberated through the Moon for more than an hour; on Earth they would have been quenched in a matter of seconds or minutes. The scientists interpreted this to mean that the Moon's outer layer consisted of loose rubble rather than solid rock. The same meteoritic bombardment that created the Moon's craters had evidently pulverized its crust, and the jangling of its pieces against each other caused the Moon to ring like a bell when struck.

As data from *Apollo 12'*s seismometer accumulated, geologists concluded that the Moon was, if not cold, at least cool, with a relatively small molten core. But the question of whether Earth's satellite had been hot at some time in its past remained unresolved. Craters caused by impacts are very difficult to distinguish from craters caused by volcanism. Perhaps some of the Moon's craters were of volcanic origin and therefore evidence of a hot past. To pursue that line of inquiry, *Apollo 13* would venture out of the flatter areas chosen as landing sites by its predecessors and head for the hills.

By now, the launch of an Apollo spacecraft was almost routine. In the tumultuous spring of 1970, with the U.S. bombing of Cambodia and student protests on campuses all around the country, the space program moved off center stage as other national concerns came to the fore. Congress had already canceled one of the ten lunar landings originally scheduled, and further cuts were being considered.

## LUCKY 13

*Apollo 13* rekindled interest in the space program, but not in any way that was intended. On April 13, as the craft was nearing the Moon, the crew of James Lovell, Fred Haise, and Jack Swigert heard a loud bang and felt the command module tremble. They immediately noticed that two of the three fuel cells in the service module had gone dead. Their first reaction was simple disappointment, because safety regulations prohibited a Moon landing with only one operating fuel cell. But as they learned more about the situation, their concern deepened: They would be lucky even to get back to Earth.

Not only were two fuel cells out of commission, but also one of the oxygen tanks in the service module had lost all of its contents, and the other was quickly venting its oxygen into space. Since fuel cells need oxygen to generate electricity, the third fuel cell would soon be dead as well. In short, the service module, ultimate supplier of oxygen, water, and electricity, was about to fail completely. It would later turn out that improper handling of the oxygen tank during manufacture had damaged its thermostatically controlled switches and wiring. During flight, the damaged switches allowed temperatures in the tank to reach 1,000 degrees Fahrenheit, burning off the wire insulation and eventually causing a short circuit. With oxygen in a gaseous state, the result was a massive explosion. The explosion damaged the other oxygen tank, leaving the command module badly crippled.

The astronauts and the ground crew immediately realized that the only dependable craft at their disposal was the lunar module. It had enough oxygen, water, and electricity to supply two people for about forty-five hours on the Moon. To get home, the crew would have to stretch those supplies to last all three members for twice as long. Since they would need the command module with its protective shield to attempt a reentry through Earth's atmosphere, the astronauts sealed off its oxygen tanks and powered down to conserve its batteries. Then the trio crawled into their lifeboat.

The first thing they had to do was establish a trajectory that would get them back to Earth as soon as possible. The service module rocket was dead, so the burn maneuvers would have to be carried out with the lunar module's descent engine. Ground controllers worked around the clock writ-

Created by TRW Systems Group, the five-rocket "mooncopter" hovers like a helicopter in this artist's conception. In theory, it could have ascended otherwise inaccessible lunar mountain peaks at speeds as great as 150 miles per hour. Designed to complement MOLAB *(opposite, top)*, the mooncopter project suffered simultaneous cancellation.

ing new procedures for this unanticipated task. But the new procedures did the trick and the firings were successful: *Apollo 13* swung around the Moon and headed for home.

During the voyage, temperatures inside the ship dipped toward freezing. To conserve water, the men drank almost nothing; the resulting dehydration caused James Lovell to lose fourteen pounds. They had trouble sleeping, and fatigue sapped both their spirits and their efficiency. Meanwhile, people around the world were following their flight.

As they neared Earth, the astronauts crawled back into the command module and powered it up with the charge remaining in its batteries. They jettisoned the service module, watching in awe as it floated off and revealed an entire side in shambles. Next to go was their lifesaving lunar module. A few minutes later, the command module alone reentered Earth's atmosphere. Ironically, the craft splashed down closer to the recovery ship than any other Apollo capsule to date—"a beautiful landing in a blue-ink ocean on a lovely, lovely planet," as Lovell later put it.

For all its drama, the perils of *Apollo 13* did not ward off further cuts in the space program. A few months later, two more lunar landings were canceled, reducing the number of remaining missions to four. Extremely frustrated, scientists and astronauts who had spent a major portion of their careers preparing for a full slate of lunar landings resolved to squeeze as much information as possible out of the remaining flights.

## OPENING UP NEW TERRITORY

One severe constraint on the spacefarers had been the amount of lunar territory they could cover on foot. *Apollo 14* brought the first wheeled device to the Moon—a rickshawlike contraption used for hauling geologic tools and rock samples. But true mobility was achieved on the *Apollo 15* mission with the debut of the lunar roving vehicle, or rover. A sort of collapsible dune buggy with a top speed of about ten miles per hour, the rover traveled in the lower section of the lunar module. When an astronaut pulled a lanyard on the side of the module, the two-seated rover popped down like a Murphy bed and its wire-mesh wheels extended. On Earth it weighed 460 pounds, but on the Moon its weight was a mere 76 pounds, so one person could lift it. With the rover, the astronauts could drive as far as six miles away from the lunar module—the most they could walk back if the wheeled vehicle broke down. Crews on the final three Apollo missions increased the area they could cover tenfold.

On *Apollos 14* through *17*, scientific exploration took absolutely top priority. All four missions landed in the Moon's light-colored highlands, a region

The 1969 Grumman rover entry came equipped with swivel seats and flexible fiberglass wheels. After a lunar mission, controllers could have operated remotely from Earth for six months. At 750 pounds, this rover was more than one and a half times as heavy as its Boeing competitor *(opposite)*.

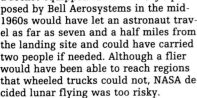

Drawing its fuel from the lunar module, a rocket-equipped lunar flier *(left)* proposed by Bell Aerosystems in the mid-1960s would have let an astronaut travel as far as seven and a half miles from the landing site and could have carried two people if needed. Although a flier would have been able to reach regions that wheeled trucks could not, NASA decided lunar flying was too risky.

of jumbled hills, sinuous valleys, and countless craters. Rock and soil samples from *Apollos 11* and *12* had indicated that the Moon's dark-colored seas took shape about a billion years after the Moon's formation. In the highlands, geologists hoped to find rocks that were as old as the Moon itself.

*Apollo 14* landed near a crater that held considerable promise. Known as Cone Crater, it had punched through the lunar surface to rock that geologists felt might be unchanged since the Moon's formation. During their second excursion, astronauts Alan Shepard and Edgar Mitchell ventured nearly a mile from the lunar module, seeking the rim of the crater. Slowed by the cart they were pulling and the steep slopes they traversed, they had to turn back before reaching the rim, though later analyses showed that they were probably within fifty feet. As it turned out, however, the rocks they collected on the way were what they had come for. The samples were more than 4.5 billion years old.

By the last Apollo flight, *Apollo 17,* the goals had come full circle: Geologist Harrison Schmitt, the first civilian scientist to come to the Moon, and his partner Gene Cernan were expressly looking for young rocks. They landed in a valley where orbital photographs from previous missions had shown the possibility of relatively recent volcanism. Finding such evidence would be a geologic coup, since it would mean that the Moon had to be hot enough to support volcanoes at some point in the not too distant past.

Near one crater, Schmitt thought he had found just such evidence. He came across soil that was distinctly orange, a common indication of volcanism. Word about the find spread quickly among earthbound geologists, and they eagerly awaited Schmitt's samples of the orange soil. But advocates of recent volcanism were to be disappointed: The orange color turned out to come from a chemically unusual form of glass more than 3.5 billion years old.

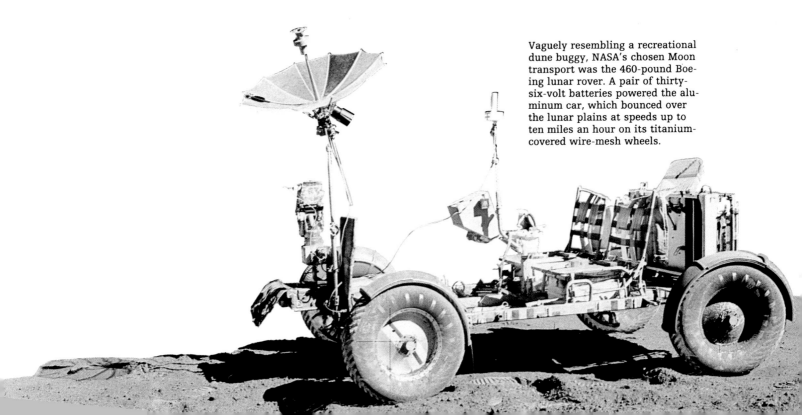

Vaguely resembling a recreational dune buggy, NASA's chosen Moon transport was the 460-pound Boeing lunar rover. A pair of thirty-six-volt batteries powered the aluminum car, which bounced over the lunar plains at speeds up to ten miles an hour on its titanium-covered wire-mesh wheels.

Although none of the six Apollo Moon landings produced any clear proof of volcanism, they completely reshaped lunar science. The rocks and soil returned from the distant world showed it to be about the same age as the Earth, approximately 4.6 billion years old. Early in its history, the Moon (and presumably Earth) underwent a period of intense bombardment, resulting in most of the craters now visible on the surface. (Craters on Earth have gradually been worn away by erosion.) Between about three billion and four billion years ago, lava from the satellite's interior flowed onto its surface and filled some of the large basins created by the bombardment, producing the dark "seas" visible today. About three billion years ago, the lava flows stopped. Except for a few relatively young craters caused by collisions, the Moon has changed little since then.

The Apollo program uncovered much about Earth's nearest neighbor, but in some respects the Moon guarded its secrets well. In particular, the data returned from the Apollo missions did not allow geologists to decide among the three leading theories of the Moon's origin: that it broke off from Earth, that it formed elsewhere in the Solar System and was captured by Earth, or that it was created as a sister planet of Earth as Earth itself took shape from the primordial dust around the Sun. Only additional expeditions to the Moon and to other planets and their satellites can settle the issue conclusively.

Landing twelve humans on the Moon was, all in all, an expensive venture, costing the United States $24 billion. But the Apollo program's achievements, and the impressions and photographs the astronauts brought back with them, are of inestimable value. Of those photographs, few have more emotional resonance than the images of a crescent Earth rising above the desolate moonscape. As astronaut Russell "Rusty" Schweickart of *Apollo 9* put it, "On that small spot, that little blue and white thing, is everything that means anything to you—all history and music and poetry and art and death and birth and love, tears, joy, and games." The sight of the Earth from space, a blue-green globe wrapped in clouds, fragile in its isolation, changed everyone who looked upon it. "You realize from that perspective," Schweickart said, "that there's something new there, that the relationship is no longer what it was."

# A SUIT FOR SURVIVAL

Beyond the blue orb of Earth lie conditions that can kill humans in any number of ways. Space contains no air to breathe. It exerts no atmospheric pressure to keep bodily fluids from boiling away. And even in the vicinity of humankind's home planet, the temperature can range from 300 degrees above zero to 300 degrees below—the difference between direct sunlight and shade.

Many of the dangers faced by astronauts venturing outside the cocoon of their spacecraft are similar to those endured by deep-sea divers, and it is no surprise that the solution should take similar form: a protective suit that, in essence, tries to replicate the nurturing environment of Earth at sea level. As illustrated on the following pages, spacesuit designers have done their best to build an outfit that not only satisfies basic needs for air, water, and atmospheric pressure but also functions as armor sturdy enough to withstand potentially lethal bombardment by microscopic meteoroids. Spacesuits must also be flexible enough to allow astronauts to manipulate tools. Finally, unlike divers, who can propel themselves by pushing against the ocean deeps, astronauts with nothing to push against need specially designed propulsion and tether systems, lest they drift aimlessly into the void.

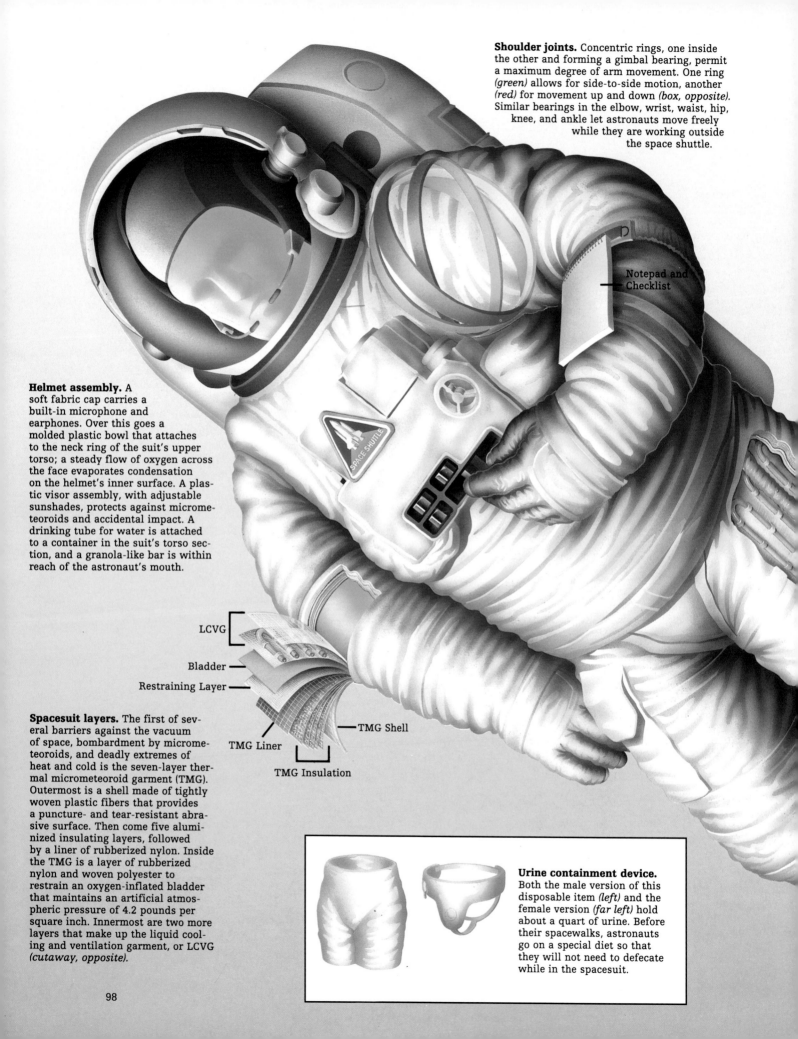

**Shoulder joints.** Concentric rings, one inside the other and forming a gimbal bearing, permit a maximum degree of arm movement. One ring *(green)* allows for side-to-side motion, another *(red)* for movement up and down *(box, opposite)*. Similar bearings in the elbow, wrist, waist, hip, knee, and ankle let astronauts move freely while they are working outside the space shuttle.

Notepad and Checklist

**Helmet assembly.** A soft fabric cap carries a built-in microphone and earphones. Over this goes a molded plastic bowl that attaches to the neck ring of the suit's upper torso; a steady flow of oxygen across the face evaporates condensation on the helmet's inner surface. A plastic visor assembly, with adjustable sunshades, protects against micrometeoroids and accidental impact. A drinking tube for water is attached to a container in the suit's torso section, and a granola-like bar is within reach of the astronaut's mouth.

LCVG

Bladder

Restraining Layer

TMG Shell

TMG Liner

TMG Insulation

**Spacesuit layers.** The first of several barriers against the vacuum of space, bombardment by micrometeoroids, and deadly extremes of heat and cold is the seven-layer thermal micrometeoroid garment (TMG). Outermost is a shell made of tightly woven plastic fibers that provides a puncture- and tear-resistant abrasive surface. Then come five aluminized insulating layers, followed by a liner of rubberized nylon. Inside the TMG is a layer of rubberized nylon and woven polyester to restrain an oxygen-inflated bladder that maintains an artificial atmospheric pressure of 4.2 pounds per square inch. Innermost are two more layers that make up the liquid cooling and ventilation garment, or LCVG *(cutaway, opposite)*.

**Urine containment device.** Both the male version of this disposable item *(left)* and the female version *(far left)* hold about a quart of urine. Before their spacewalks, astronauts go on a special diet so that they will not need to defecate while in the spacesuit.

# ANATOMY OF AN ENGINEERING MARVEL

The suit worn by space shuttle crew members during extravehicular activities, or EVAs, reflects efforts to increase an astronaut's mobility without sacrificing vital protection. The suit's tough but pliable outer skin—a complex layering of synthetic materials—acts like a bulletproof vest to stop dust-size micrometeoroids that would easily penetrate conventional fabrics. Oxygen pumped into an inner lining of the suit maintains a constant pressure, preventing the astronaut's blood and other body fluids from boiling in the vacuum of space.

Most of the suit's components are available in a range of sizes from extra large to extra small, enabling astronauts to assemble pieces to fit. Suits are stored in an air lock that also serves as a dressing room. After donning a urine containment device *(box, opposite)* and a skin-hugging garment designed for cooling and ventilation, the astronaut climbs into the pants and boots, which come as a single piece. This lower unit joins at the waist to the upper torso, a vestlike shell of molded fiberglass with flexible arms and gloves and a control and instrument panel mounted on the front. The astronaut slides up into the torso, which is attached to the life-support backpack *(pages 100-101)*, then puts on the helmet assembly. On Earth, the spacesuit and life-support pack would together weigh approximately 250 pounds and would be an ordeal to get into. In the zero gravity of space, dressing and then checking the suit for safety takes all of twenty minutes.

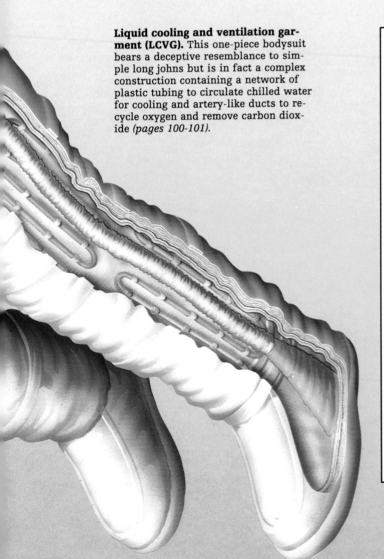

**Liquid cooling and ventilation garment (LCVG).** This one-piece bodysuit bears a deceptive resemblance to simple long johns but is in fact a complex construction containing a network of plastic tubing to circulate chilled water for cooling and artery-like ducts to recycle oxygen and remove carbon dioxide *(pages 100-101)*.

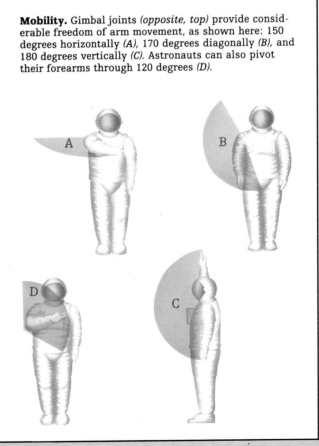

**Mobility.** Gimbal joints *(opposite, top)* provide considerable freedom of arm movement, as shown here: 150 degrees horizontally *(A)*, 170 degrees diagonally *(B)*, and 180 degrees vertically *(C)*. Astronauts can also pivot their forearms through 120 degrees *(D)*.

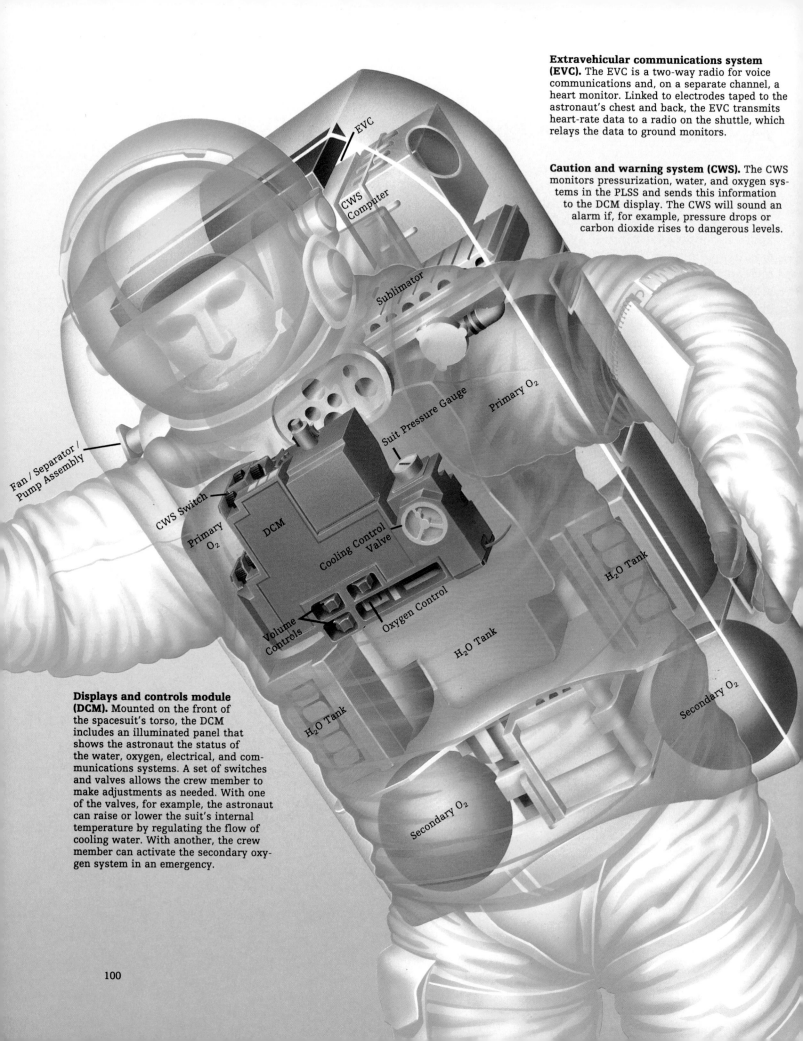

**Extravehicular communications system (EVC).** The EVC is a two-way radio for voice communications and, on a separate channel, a heart monitor. Linked to electrodes taped to the astronaut's chest and back, the EVC transmits heart-rate data to a radio on the shuttle, which relays the data to ground monitors.

**Caution and warning system (CWS).** The CWS monitors pressurization, water, and oxygen systems in the PLSS and sends this information to the DCM display. The CWS will sound an alarm if, for example, pressure drops or carbon dioxide rises to dangerous levels.

**Displays and controls module (DCM).** Mounted on the front of the spacesuit's torso, the DCM includes an illuminated panel that shows the astronaut the status of the water, oxygen, electrical, and communications systems. A set of switches and valves allows the crew member to make adjustments as needed. With one of the valves, for example, the astronaut can raise or lower the suit's internal temperature by regulating the flow of cooling water. With another, the crew member can activate the secondary oxygen system in an emergency.

EVC

CWS Computer

Sublimator

Fan / Separator / Pump Assembly

Suit Pressure Gauge

Primary $O_2$

CWS Switch

Primary $O_2$

DCM

Cooling Control Valve

$H_2O$ Tank

Volume Controls

Oxygen Control

$H_2O$ Tank

$H_2O$ Tank

Secondary $O_2$

Secondary $O_2$

# An Astronaut's Link to Life

Working outside the space shuttle would not be possible without a vital item of gear known as the primary life-support subsystem, or PLSS. Bolted to the back of the spacesuit's fiberglass torso, the PLSS (pronounced "pliss") circulates a continuous flow of water and oxygen for cooling, ventilation, and breathing (below)—enough to let a crew member work for six and a half hours at a level of exertion equivalent to pedaling a bicycle at twelve miles per hour. As illustrated at left, a displays and controls module—the instrument panel for the PLSS—is mounted on the chest. At the top and bottom of the backpack itself are, respectively, a two-way radio and a secondary oxygen pack that can supply thirty minutes of emergency oxygen should the main circuit fail.

The primary system is designed to allow astronauts to breathe pure oxygen at a pressure of 4.2 pounds per square inch. Because this is considerably less than the Earth-like atmospheric pressure of 14.7 psi in the space shuttle cabin, crew members must get acclimated to avoid the sudden drop in pressure that causes the bends. Accordingly, some thirteen hours prior to a spacewalk, air pressure in the cabin is reduced to 10.2 psi. The EVA crew then breathes pure oxygen for an hour. About an hour before EVA, the crew suits up in the air lock and breathes pure oxygen for another forty minutes. Pressure in the air lock is then reduced to 5 psi as the astronauts run through a last check on their suits. Finally, pressure drops to 0 psi, and the air lock opens to the vacuum.

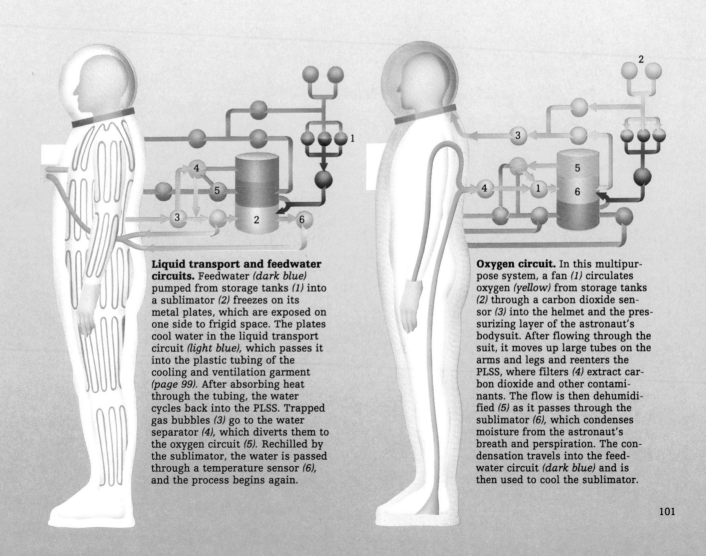

**Liquid transport and feedwater circuits.** Feedwater (dark blue) pumped from storage tanks (1) into a sublimator (2) freezes on its metal plates, which are exposed on one side to frigid space. The plates cool water in the liquid transport circuit (light blue), which passes it into the plastic tubing of the cooling and ventilation garment (page 99). After absorbing heat through the tubing, the water cycles back into the PLSS. Trapped gas bubbles (3) go to the water separator (4), which diverts them to the oxygen circuit (5). Rechilled by the sublimator, the water is passed through a temperature sensor (6), and the process begins again.

**Oxygen circuit.** In this multipurpose system, a fan (1) circulates oxygen (yellow) from storage tanks (2) through a carbon dioxide sensor (3) into the helmet and the pressurizing layer of the astronaut's bodysuit. After flowing through the suit, it moves up large tubes on the arms and legs and reenters the PLSS, where filters (4) extract carbon dioxide and other contaminants. The flow is then dehumidified (5) as it passes through the sublimator (6), which condenses moisture from the astronaut's breath and perspiration. The condensation travels into the feedwater circuit (dark blue) and is then used to cool the sublimator.

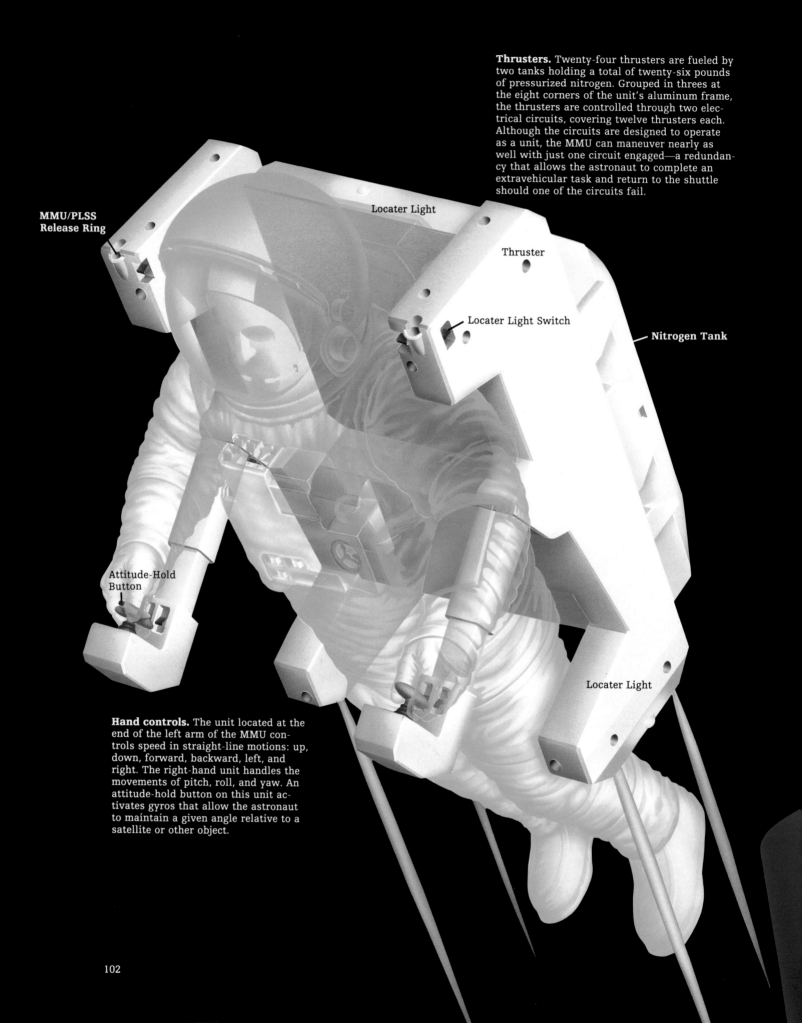

**Thrusters.** Twenty-four thrusters are fueled by two tanks holding a total of twenty-six pounds of pressurized nitrogen. Grouped in threes at the eight corners of the unit's aluminum frame, the thrusters are controlled through two electrical circuits, covering twelve thrusters each. Although the circuits are designed to operate as a unit, the MMU can maneuver nearly as well with just one circuit engaged—a redundancy that allows the astronaut to complete an extravehicular task and return to the shuttle should one of the circuits fail.

MMU/PLSS
Release Ring

Locater Light

Thruster

Locater Light Switch

**Nitrogen Tank**

Attitude-Hold
Button

Locater Light

**Hand controls.** The unit located at the end of the left arm of the MMU controls speed in straight-line motions: up, down, forward, backward, left, and right. The right-hand unit handles the movements of pitch, roll, and yaw. An attitude-hold button on this unit activates gyros that allow the astronaut to maintain a given angle relative to a satellite or other object.

# A Personal Power Pack

Once out of the air lock, astronauts are faced with the tricky prospect of moving themselves around in zero gravity. If the EVA assignment is within the confines of the shuttle's cargo bay—repairing the remote manipulator arm, for example—crew members can make use of retractable tethers and foot restraints to keep themselves from drifting away.

To navigate the emptiness between the shuttle and a satellite in need of repair or retrieval, however, astronauts require another form of propulsion. The manned maneuvering unit (MMU) is a twenty-four-thruster jet pack powered by pressurized nitrogen gas. With the MMU snapped into place on the life-support

backpack *(opposite),* space walkers could travel as fast as sixty feet per second (more than forty miles per hour), but in reality they never operate the unit at more than a fraction of that. Experiments have shown that an astronaut flying toward an object in space feels most in control of the MMU at speeds of about 1 foot per second for every 100 feet of distance to the object. In approaching a satellite that is 300 feet away, for example, a crew member would move at about 3 feet per second—roughly the same pace as a stroll on Earth. The MMU enables its pilot to match the spin rate of a tumbling satellite, dock with it, and stabilize it using counterthrusts from the MMU jets.

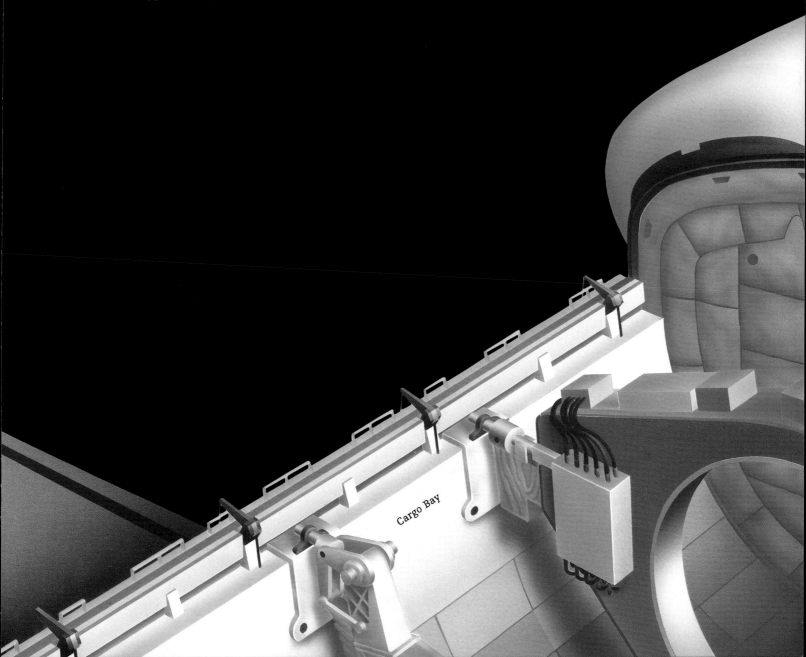

Cargo Bay

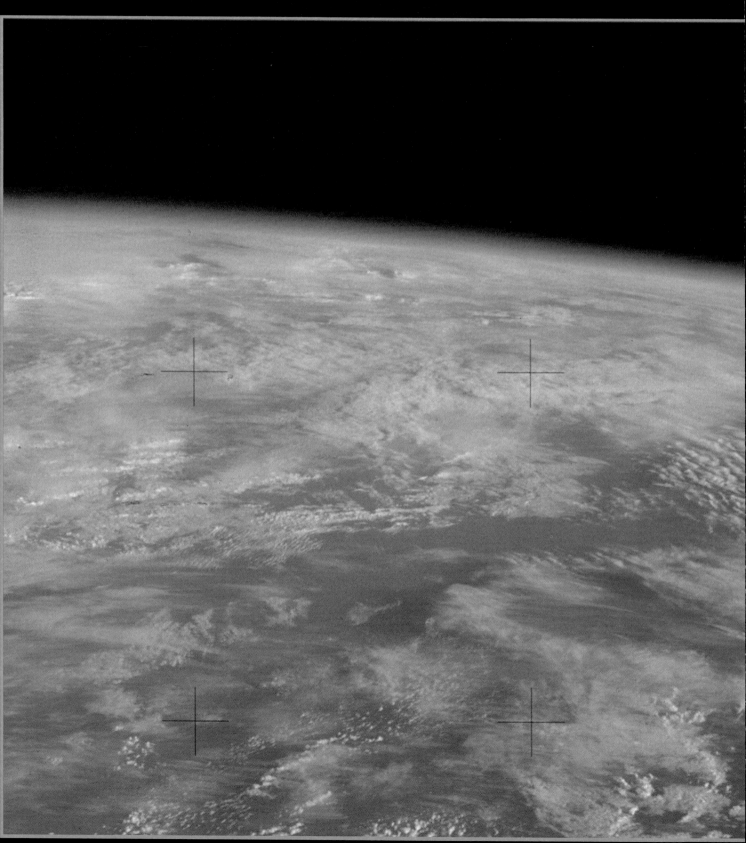

ike some leviathan stirring in the deeps, Earth turned slowly below the space station, its curving edge opalescent against the blackness of space. Inside *Salyut 7*, the crew had attempted to re-create a bit of the ambiance of their home world: The cabin was warmly lit, brightly colored, and domestic; a tidy setting of table and chairs perched next to a kitchen packed with pots and plates. Yet the cosmonauts of the 1982 mission struggled with a disturbing sense of isolation, a feeling accentuated, as flight engineer Valentin Lebedev put it, by the "dead silence all around."

During their 211-day flight, cosmonauts Lebedev and Anatoli Berezovoy grew increasingly hungry for the small sensory details that meant Earth. Lebedev, a businesslike spacecraft designer, found himself drawn to the tiny, tenderly nurtured space garden where he coaxed along a variety of sprouts: peas, wheat, oats, parsley, onions, dill, fennel, and garlic. In the midst of some of the most technologically advanced hardware, with a rare and magnificent view always in front of him, the engineer discovered that his greatest excitement lay in caring for the young plants. When a tiny leaf unfurled, he said later, "it seemed to fling open a bright window out into the world."

The new leaves represented not only the planet left behind but also the challenge of achieving self-sufficiency in space. Cut off from the bounty of mother Earth, with its generous supplies of oxygen and water and its teeming, self-propagating life, both Soviet and American spacefarers have attempted to construct their own imitations of a living world within orbiting spacecraft. When and if they succeed, they will be ready for the next step: long voyages of interplanetary exploration.

As they look toward longer stays in space, the two countries have been trying to solve three major problems. The first is the need to provide enough food and oxygen to sustain a crew away from home. Soviet experts calculate that a three-person crew on a one-year mission needs 1.5 tons of food, 3.3 tons of oxygen, and 5.9 tons of water; those supplies alone total almost half the weight of a current Soviet space station. The second problem is physiological: Doctors know that a zero-gravity environment affects the human body in a variety of ways, short-circuiting blood production and causing muscles to atrophy and bones to thin. Throughout the 1970s and 1980s, physicians tested the blood, measured the skeletons, and supervised the diets of cosmonauts and astronauts, trying to understand these effects and learn how to combat them. Less measurable, and possibly more dangerous, is the third issue, isolation psychology. Will the enforced company of a few other people for months on end erode the mental health of a spacecraft's crew?

## SOVIET TRAILBLAZING

The Soviet Union tackled these issues directly when it took the lead in developing space stations late in the 1960s, even as the United States surged ahead in the race to the Moon. While the Americans pursued the Apollo program, the Soviets put up their own series of Soyuz vehicles. The insectlike Soyuz craft, fitted with two solar panel wings, could hold three people and was notable for the long docking probe mounted like a stinger on its forward end. By January of 1969, two of the vehicles, *Soyuz 4* and *Soyuz 5*, had accomplished the first docking between piloted craft, six weeks before *Apollo 9* managed a similar feat. The Soviets called the joined vehicles "the world's first experimental assembled cosmic station."

In June of 1970, *Soyuz 9* set a different kind of record by supporting its two crew members for an unprecedented eighteen-day stint. In what would become a familiar exchange between ground control and space dwellers, Soviet command chided the cosmonauts for not exercising more, to which they replied irritably that they did not have the time. Perhaps as a result, the two pioneers had to be carried from their reentry capsule when they landed.

Although the Soviets may have intended to head for the Moon earlier in the Soyuz program, the achievements of the *Soyuz 4, 5,* and *9* missions made it clear to Western scientists that they now had another goal: establishing a workable space station. And within a year, in April 1971, they brought forth *Salyut 1.* The trailblazing craft would turn out to be singularly ill fated.

*Salyut 1* was an uncomplicated metallic cylinder about sixty-five feet long and thirteen feet wide. Launched without crew atop a medium-size booster rocket, the world's first operational space station was divided into several compartments. Three were pressurized: the principal module where the crew would live, the equipment module containing supplies and controls, and the access module, which cosmonauts would enter from their docked spacecraft.

The goal of the earliest missions to the station was simply to get on board through the single docking port. The first crew sent to *Salyut 1* aboard *Soyuz 10* docked successfully—and then, to the bafflement of Western ground observers, remained hooked up for less than six hours, never entering the station at all. American space experts later speculated that some kind of technical problem prevented entry, but the Soviets never released an explanation.

In June 1971, a second crew, aboard *Soyuz 11,* returned to Salyut. The commander was Lieutenant Colonel Georgi Dobrovolsky, a tough fighter pilot and long-time cosmonaut; he was accompanied by two civilians, flight engineer Vladislav Volkov and Victor Patsayev, who had been a radio researcher and engineer until he joined the space program. The *Soyuz 11* team performed a flawless docking, entered the space station, and began at once to spruce up their new quarters. The exuberant threesome became Soviet television stars

## ACHIEVING A HANDSHAKE IN ORBIT

It sounded so simple in 1972, when President Richard Nixon and Premier Alexei Kosygin signed the agreement: a joint United States-Soviet Union mission that would be capped by a rendezvous in orbit. But preparing for a hookup of Apollo *(above, left)* and Soyuz *(above, right)* proved to be a three-year effort. Astronauts and cosmonauts had to overcome the language barrier to coordinate their different flight techniques. Bilateral teams of scientists and flight engineers faced similar hurdles in inventing a common docking module for Apollo's pin-and-cone design and the Soyuz flower-petal system. The docking module was also the crucial link between two radically divergent life-support systems. The atmosphere aboard the Soyuz was a mix of oxygen and nitrogen, kept at 14.7 pounds per square inch (psi), the normal atmospheric pressure at sea level. The Apollo's interior, in contrast, was 100 percent oxygen at 5 psi. Linkup under such different pressurizations would give the Soyuz crew a nasty, and perhaps fatal, case of the bends. On July 15, 1975, the two craft were launched, seven and a half hours apart. Two days later, as they bumped together 140 miles above Metz, France, the atmosphere in the Soyuz was reduced to 10 psi. The Apollo crew, which had carried the docking module into orbit, raised the module's atmospheric pressure to the same mutually tolerable level. Then commanders Thomas Stafford and Alexei Leonov crawled toward each other to meet in the module's padded hatchway for their historic handshake.

almost immediately, somersaulting and joking before the station's cameras. They also demonstrated the usefulness of a space-based science lab, surveying Earth's land and sea and observing stars without the interference of an atmosphere. Their favorite task, though, was working with the pint-size menagerie brought on board. "Our pets give us great pleasure to watch," they said of their collection of tadpoles and fruit flies.

After twenty-three days, the *Soyuz 11* crew left the space station and set off for Earth, ready for a triumphant homecoming. But their descent through the atmosphere occurred in eerie silence. When ground crews pried open the capsule after a perfect landing, they found all three men still strapped into their seats, dead. Later, the Soviets acknowledged that as the spacecraft separated from the station, vibrations from the explosive bolts that detached the command module from Salyut must have weakened a valve on the cabin, allowing the precious air to leak slowly away. The crew, crowded within the Soyuz capsule, had not been wearing pressurized suits. Never again would cosmonauts fly without that protection. Nor would they ever fly again to *Salyut 1.* Several months later, the Soviets slowed the orbit of the unlucky station, deliberately bringing it down over the Pacific Ocean.

Eighteen months passed before the program cautiously resumed, only to

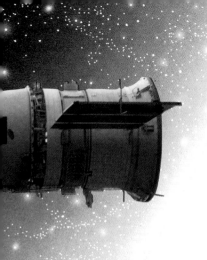

meet with more setbacks as two subsequent stations, *Salyut 2* and *Kosmos 557*, broke into pieces before reaching orbit. Finally, in June 1974, *Salyut 3* restored Soviet pride with a cloak-and-dagger mission. The station hovered in a lower orbit than usual, employing military crews who corresponded with the ground over secret communications channels. *Salyut 5*, launched two years later, was also apparently used for intelligence activities. But between those two was *Salyut 4*, launched in December 1974, which housed two successful missions over six months. The cosmonauts aboard tested a new space exercise bike and performed some simple biological experiments, such as growing bacterial cultures and raising fruit flies.

Perhaps more important, the early Salyuts spawned a new generation of space stations: *Salyuts 6* and *7*, followed by the even more sophisticated Mir. The new Salyuts, the first of which was launched in September 1977, differed from their predecessors by the addition of a second docking port, allowing resupply flights to unload without the crew first undocking their own spacecraft to let the supply ship park. Shortly afterward, the Soviets began to resupply and refuel the station using an automated cargo ship.

Most notably, these space stations began to host crews for increasingly lengthy missions. Cosmonauts Georgi Grechko and Yuri Romanenko set a record when they spent ninety-six days aboard *Salyut 6* in 1977 and 1978. Their stay almost ended in tragedy, though, when a curious Romanenko took a peek at space through an open hatch while his colleague was doing some outside repairs. Unfortunately, Romanenko had forgotten to tether his pressurized suit. Grechko, working alongside the hatch, was alarmed to see his companion suddenly float out, his arms clutching at emptiness as he desperately tried to stop himself. Grechko grabbed the safety line as Romanenko drifted by, and the two shaken men managed to return to safety.

Their mission opened the way for even longer visits. *Salyut 7*, launched in 1982, held a crew of three for 237 days in 1984. Although that mission was a triumph, the station itself was not. *Salyut 7* developed so many electrical power problems that almost every crew had to take spacewalks to repair it. Soon *Salyut 7* was supplanted by a grander vehicle—the third-generation space station Mir, lofted into orbit in 1986.

Mir was a great improvement over its predecessors. Roomier and more versatile, the fifty-six-foot-long station had six docking ports and a research laboratory within a separate module. The additional docking points helped alleviate crowding in the main station; scientific equipment could be installed on separate modules attached to the ports. Mir also demonstrated advances in recycling, a critical matter if space stations are to succeed. Special condensers were designed to trap the 4.5 pounds of water vapor released every twenty-four hours by each cosmonaut's breath and sweat and convert at least some of it back into usable liquid, eliminating the need to transport so much water. Mir's eight on-board computers regulated this function and others, freeing the cosmonauts from tedious chores. The space station's improved recycling technology and advanced computer control made it a more comfortable home for cosmonauts, and in late 1988 a crew returned to Earth after spending an unprecedented 366 days aboard.

From 1971 through 1988, Soviet space stations hosted more than fifty men and one woman for a total of over seven years of weightless living. In contrast, the United States launched one station, Skylab, in 1973; nine people crewed the floating laboratory for a total of 171 days. As a result, research in zero-gravity life has fallen largely to the Soviets. Their cosmonauts have found that space-station living poses some substantial physical and psychological challenges.

## MICROGRAVITY

Perhaps the most alien feature of space-station life is weightlessness. Except for the absence of gravity, occupying a space station would be little different from living in the cramped quarters of an oceangoing submarine. The environment of the space station is not, in fact, quite gravity free. The bulky structure with its humming engines generates just a trace of gravitational force—about one ten-thousandth of a g—creating microgravity, as physicists call it. However, the average person cannot feel such a minuscule force, and in many ways space-station crews have

**1935** Drägerwerke, a German diving-suit manufacturer, fashioned a suit of silk and rubber covered with silk cords. When inflated, however, the suit expanded so much that the eyepieces rose up onto the pilot's forehead.

# OUTERWEAR FOR OUTER SPACE

The modern spacesuit is descended from pressurized garb designed for deep-sea divers and high-altitude aviators. Many of these ancestral suits shared a serious disadvantage: They were virtually impossible to move around in. Either the suits were too stiff or bulky to begin with, or they became so once they were pressurized. The garments took two basic forms: "soft" suits, with a certain amount of flexibility but less than full protection against tears and punctures, and armorlike "hard" suits, safer but extremely ponderous. Most high-altitude aviators opted for soft suits simply because they were easier to wear. But true space flight has tended to favor balanced solutions. Today, hard suits with better flexibility have become the outer-space outerwear of choice.

**1933** In a suit designed by British engineers John Scott Haldane and Sir Robert Davis, American Mark Ridge survived in a chamber simulating atmospheric pressure at an altitude of 90,000 feet.

**1934** Pilot Wiley Post (above) and B. F. Goodrich designer Russell Colley developed a suit similar to a deep-sea diver's. One earlier design ruptured; another had to be removed by cutting it off.

**1935** This French-designed outfit of linen, silk, and rubber came equipped with spring-loaded gloves that let the pilot turn a chest crank to control the garment's internal pressure.

found weightlessness immensely liberating. Although the décor of most stations imposed a traditional up-and-down orientation, the occupants frequently ignored it. Aboard Skylab, for example, crews indulged in riotous horseplay at times, bouncing off the walls and flying paper airplanes. They found that they could sleep in almost any position: Pete Conrad, the cheerful commander of the first Skylab crew, chose to sleep hanging from his feet because he did not like the way air from the ventilation system blew into his nose if he was in an "upright" position. To maneuver in zero gravity, crew members used hooks, straps, adhesive fasteners, and boots that clamped onto grids in the floor. Without these aids, station dwellers found themselves oddly helpless: When they stalled in the center of the cabin with nothing to grab, they would hang suspended in air until a colleague gave them a shove.

Cosmonauts also enjoyed the acrobatic potential of weightlessness. In his diary of the 1982 *Salyut 7* stay, Soviet engineer Lebedev noted that he and his companion swam from point to point like airborne amphibians. In order to sit down without floating away, the cosmonauts had to place themselves backward in chairs, wrapping their arms and legs around the chair

**Late 1930s** The first hard suit, by an Italian armor maker, proved too bulky and heavy. After squeezing through a rear entry, hapless pilots had to be carried to their planes.

**1940** A rubberized British suit, outfitted with a thigh-mounted emergency oxygen bottle and adjustable leg lacings, performed well at an altitude of 35,000 feet.

**1940** The first American spacesuit, commissioned by the Army Air Forces, weighed 80 pounds and became completely rigid at 3 pounds per square inch, 0.5 psi short of the pressure needed to sustain human life in space.

**1942** An experimental series of Army Air Forces suits included the unsuccessful XH-1. Among other drawbacks, it had poor ventilation, and it became unmanageable when it was inflated.

**1942** Reminiscent of medieval armor, a hard suit of German design proved more flexible than most and could withstand high internal pressures. But like other hard suits, it was too heavy.

backs. "Sometimes our body positions become so outrageous, even monkeys couldn't compete with us," he said.

The gravity-free environment, however, can be as disturbing as it is exhilarating. Space sickness, the nausea experienced by some astronauts and cosmonauts from the beginning of the space program, became a major problem in space stations. At first, Skylab's visiting crews were thrilled about their larger environs; in comparison with the tiny space capsules of the 1960s, the new compartments were palatial, featuring private sleeping quarters, kitchens, exercise rooms, and research sections. But along with the added elbow-room came dizziness and nausea. Researchers at NASA realized that astronauts on Mercury, Gemini, and even some Apollo flights were so tightly wedged into place that they rarely jostled the sensitive fluids of the inner ear that control balance. In the roomy space stations, however, the floating sensation rapidly became sickening. Fortunately, the effect was usually temporary, and orbiting crews eventually adjusted, sometimes with the help of medication.

Space sickness is not the only unpleasant effect of microgravity. With the onset of weightlessness, many cosmonauts reported an unsettling feeling that they were tipping over

**1943** Inspired by the segmented body of a worm, inventor Russell Colley devised the "tomato-worm suit." Its jointed construction let wearers lift and bend their limbs.

**Early 1950s** The David Clark Company, a maker of pressure suits since 1939, branched out into brassiere manufacture before developing this garment made of nylon waffle-weave fabric.

**1960** This Soviet pressure suit appeared within months of the capture of American U-2 pilot Francis Gary Powers, who was shot down over the Ural Mountains in a similar high-altitude suit.

**1962** Although it restricted visibility and was somewhat rigid even at low pressures, this British suit was lightweight and introduced the zippered rear entry, later used in Apollo spacesuits.

backward, followed by a headachy sensation of blood rushing to the face. This congestion occurs when bodily fluids float up to the head, leaving the face puffy and flushed and the eyes slanted. "Our weightlessness isn't that much of a pleasure," noted one disgruntled Salyut cosmonaut. "Our faces have begun to swell, so much that looking into the mirror I fail to recognize myself. I keep bumping into things, mostly with my head." Other physical effects show up more gradually. Heart rates slow in the weightless body; blood production drops, bones thin, and appetites vanish. Without the stress of working against gravity, leg muscles begin to atrophy.

Medical researchers monitoring the orbiting teams still do not know how permanent some of these effects might be. Scientists studying cosmonauts on missions approaching a year in length have found that the travelers' bones lose about 0.5 percent of their calcium per month. The loss in weight-bearing bones, such as the heel and the shin, has reached as high as 5 percent. There is considerable scientific debate about whether the mineral depletion of bones levels off—and whether the loss is recoverable.

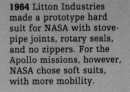

**1964** Litton Industries made a prototype hard suit for NASA with stovepipe joints, rotary seals, and no zippers. For the Apollo missions, however, NASA chose soft suits, with more mobility.

**1963** The Mercury spacesuit, shown here on Gordon Cooper, was made of aluminum-coated nylon and rubber, weighed only twenty pounds, and had better ventilation than that of any previous suit.

**1962-1964** This experimental spacesuit helped the American space program develop fabrics that were resistant to radiation and micrometeoroid penetration.

**1965** Made by the David Clark Company and worn during the first American spacewalk, the Gemini spacesuit had an antiglare gold-coated visor and directed the astronaut's exhalations away from nose and mouth.

113

Soviet researchers are more confident of their ability to control muscle degeneration. After about a month in space, crew members generally lose around 20 percent of the strength in their leg muscles and 10 percent in their arms, but the muscles reassert themselves upon return to Earth. Nevertheless, both the United States and the Soviet Union have insisted on vigorous physical fitness programs. Cosmonauts, for instance, have been required to exercise at least two hours a day—and have not enjoyed it. Salyut cosmonaut Valeri Ryumin noted, "I hate our exercises. Loved it on Earth. But here, each time I have to force myself. Boring and monotonous, and heavy work."

During the ninety-six-day *Salyut 6* mission, the crew was required to work out on a treadmill wearing a harness, on an exercise bike, and with bungee cords. The regimen was exhausting; two hours of it was the equivalent of climbing a 200-story skyscraper. Grechko and Romanenko tended to put off the workouts as often as possible, but when they returned to Earth on March 16, 1978, they had to be carried from the capsule, and tests showed their heart volume had diminished and calf muscles shrunk. Following the 237-day *Salyut 7* mission in 1984, crew members spoke vividly of the crushing weight that greeted them back on Earth. As flight engineer Vladimir Solovyov, a civilian space-station designer, put it, "I wake up in the morning and my first

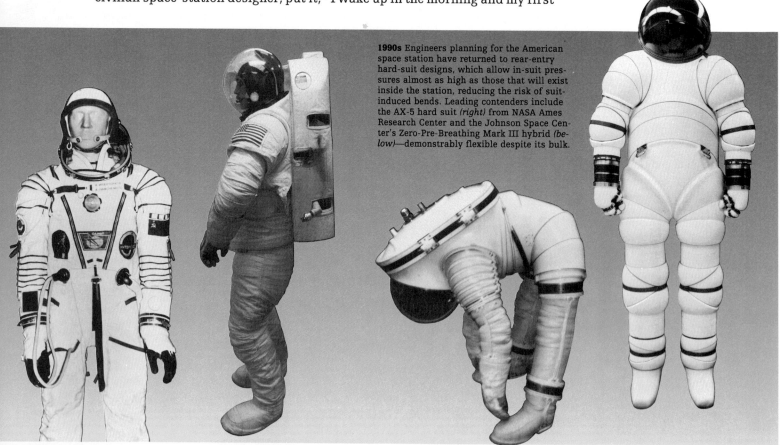

**1990s** Engineers planning for the American space station have returned to rear-entry hard-suit designs, which allow in-suit pressures almost as high as those that will exist inside the station, reducing the risk of suit-induced bends. Leading contenders include the AX-5 hard suit *(right)* from NASA Ames Research Center and the Johnson Space Center's Zero-Pre-Breathing Mark III hybrid *(below)*—demonstrably flexible despite its bulk.

**1970s-1980s** Cosmonauts wore soft suits similar to this one on most of the Salyut space-station missions. More recently, the Soviets have favored a hard-suit design with a rear-entry hatch.

**1980s** American shuttle astronauts wear a hybrid suit that combines hard and soft materials, thus providing both mobility and security. Modular parts allow more individualized fitting.

thought is, 'Why didn't I break the bed?' " Nor has it been easy for cosmonauts to adjust mentally; for a while after their return, Grechko and Romanenko tried to swim out of their beds every morning.

Although exercise clearly helps to stave off the ill effects of living without gravity, doctors cannot yet decide whether very long periods of weightlessness, such as would be sustained during a two- to three-year trip to Mars, would irretrievably damage the human body. Further, they cannot be confident of the ability of even well-adjusted people to cope with the double burden of claustrophobia and isolation that comes with life aboard a spacecraft. Space-station psychology has proved to be a tricky business.

**CABIN FEVER**

From the start, U.S. and Soviet engineers knew they had to give their orbiting guinea pigs as much room as possible to stave off cabin fever. Skylab, for instance, was built from the empty upper stage of a massive Saturn V rocket. The entire assembly was 117 feet long and weighed nearly ten tons. The interior was split into two floors separated by a metal grid: The upper level served as a warehouse, and the lower contained workspace, three tiny bedrooms, a bathroom, and a living-dining-kitchen area called the wardroom.

Plans for the wardroom established one of the guiding principles of space stations: People in orbit get to see out. During the early NASA design sessions, however, the window was a point of controversy. In an argument reminiscent of the early Mercury program, the Skylab director in Huntsville, Alabama, felt that the space station should be designed for efficiency and that a window was an expensive frill. But the crew could not imagine floating over Earth packed away like beans in a can. Finally, the agency scheduled a meeting at its Washington, D.C., headquarters to discuss station livability. Included in the session was Raymond Loewy, a top industrial designer and the creator of the elegant 1953 Studebaker. He insisted that a window was essential. So a window was added, much to the delight of astronauts Pete Conrad, Paul Weitz, and Joe Kerwin, who went on to spend twenty-eight days in space, followed by two other teams in missions of fifty-nine and eighty-four days.

Soviet engineers also provided for windows aboard the Salyut stations, but the most advanced example of engineering psychology so far is Mir. Heeding the comments of earlier Salyut crews, designers provided as much privacy as possible in the modern station; each cosmonaut has his or her own small cabin with a porthole, sleeping bag, desk, and armchair. The station's interior has been decorated in soft, presumably soothing colors—yellow, light green, and tan. Cosmonauts cook on a hot plate and select their own foods, as long as they consume the required 3,200 calories a day.

Despite its homey qualities, a station like Mir is also a prison, albeit a voluntary one. Space-station crews are screened for compatibility and tolerance of isolation, but experience has shown that irritability levels tend to rise with every week of a mission. During the Salyut missions, ground controllers noticed that cosmonauts grew increasingly testy at what they deemed stupid questions and became strangely secretive, even holding back information. In response, controllers inaugurated strict communication rules: Among other things, they avoided negative remarks and divided their time equally among all cosmonauts. During the extended 1977 mission of Grechko and Romanenko, ground controllers also established a "psychological support group" to monitor the cosmonauts' mental health using voice-stress analysis—a technique that measured the ease, tempo, and response time of the station dwellers' speech. Specialists processing the data sent by cosmonauts to Earth also reported back regularly, allowing the orbiting travelers to see how valuable their work was.

In addition, the Soviets bombarded the crew with care packages. Every month supply ships brought up letters and gifts, once a guitar, and another time a package of colored partitions to increase privacy. The men themselves, naturally aggressive and ambitious, worked hard to avoid trying to outdo one another. "Competition within a crew is one of the most harmful things," Grechko said later. "Especially if each starts trying to prove that he is the best one." In space, the cosmonaut noted, "you have no psychological outlets. It is much more dangerous there."

Like their Soviet counterparts, the U.S. controllers who monitored Skylab also learned not to push the isolated crews too much. The final team, headed by Jerry Carr, a lieutenant colonel in the Marines, entered the station on November 16, 1973, for an eighty-four-day mission. The three members were loaded down with work, including the task of observing the Sun, photographing an active flare from its corona and recording its ultraviolet and x-ray spectrum, regions of radiation that cannot penetrate Earth's atmosphere. Within a few weeks, Carr's group began complaining bitterly about not being given time to just look around and enjoy the silence of space. About a third of the way into the mission, astronaut Edward Gibson lamented, "I personally have found the time since we've been up here to be nothing but a thirty-three-day fire drill." In the sixth week, Carr, Gibson, and William Pogue simply went on a one-day strike. Gibson, a solar physicist, spent his time watching the Sun. The other two used the day to take a long look at Earth. As a result, the crew reached a compromise with mission control that gave them more dominion over their own time; later, in fact, they picked up the pace of the experiments. "When I tried to operate like a machine, I was a gross failure. We've got to appreciate a human being for what he is," Pogue said.

Carr's group was the last Skylab crew. Although NASA had planned to send its space shuttle to the station, Skylab tumbled out of orbit on July 12, 1979, a year and nine months before the first shuttle flew. Scientists later calculated that a period of high solar activity warmed the Earth's atmosphere, expanding it toward the station's nearly 300-mile-high orbit and dragging the craft down with increased friction.

Despite the Skylab revolt, most space-station crews reported that work was the best antidote to restlessness. Scientific tests aboard the various stations ranged from the charming—a 1979 effort to see how tulips fare in the absence

With eyes glued to individual consoles, overhead panels, and large screens, controllers at Johnson Space Center in Houston monitor the fourth flight of the shuttle *Challenger* in February 1984. Information on the flight—including a live view of the cargo bay *(center screen)*, where an astronaut practices the capture of a satellite—was relayed to Houston through tracking stations around the world. In the foreground, three controllers designated capsule communicators, or "CapComs," relay instructions to the orbiting crew from flight director John Cox *(far left)*.

of gravity—to more elaborate attempts to create new polymers in weightless conditions. Biological experiments consisted of both research into the effects of weightlessness and endeavors to promote small, self-sustaining farms. Skylab astronauts tested the ability of two spiders—Anita and Arabella—to spin webs in microgravity and noted with amusement the crooked formations that resulted before the dizzy arachnids adapted. They found that first-generation minnows were also disoriented, but their offspring navigated easily in the plastic tanks, simply choosing one wall as their "floor."

The Soviets carried aloft amebas, viruses, rats, monkeys, guppies, flies, and other insects to see how a variety of life forms survive in space. They also experimented with artificial gravity by placing test animals inside a small centrifuge. The experiments have pointed the way to one possible solution to the debilitating effects of weightlessness: In rats, at least, the simulated gravity maintained bone and muscle mass, while keeping heart, blood, and hormone functions normal. The cosmonauts have struggled less successfully with plants. Some small crops, such as peas and lettuce, have sprouted into acceptable food. But more often the plants have grown normally for a brief period and then died. Orchids, brought aboard *Salyut 6,* promptly dropped their flowers and refused to bud again until they were returned to Earth.

Nevertheless, Soviet planners have pushed rapidly forward with further experiments and longer missions, building up the expertise that they hope will take them to Mars by the year 2010. Meanwhile, the United States has

followed its own path. While the Soviets were fine-tuning the art of living in space, the Americans developed another essential element of a future program of exploration: the space ferry.

## BRIGHT HOPES FOR THE SHUTTLE

Long before Skylab was boosted into orbit, NASA officials dreamed of serving their space station with a spacecraft that would take off like a rocket and land like an airplane. In 1967, the president's Science Advisory Committee began to plan for both a space station to succeed Skylab and a means of shuttling between it and Earth. As proposed by NASA engineers at the time, the shuttle would be a two-stage, totally reusable vehicle. The first stage, a bulky, winged craft with a crew of two, would provide the initial rocket-powered thrust out of the atmosphere and then return to a runway. The second stage, released into orbit, would also have two pilots and could ferry up to twelve passengers to a space station before returning in a similar airplane-style landing. But by 1970, public enthusiasm for federal programs was foundering in the Vietnam War controversy, so President Nixon decided on a more modest approach: a stubby-winged, bullet-nosed orbiter.

As the shuttle evolved during the 1970s, demands by military and civilian customers helped shape it into a satellite-carrying workhorse. Astronauts described the final product as a "flying machine of unparalleled complexity." The 122-foot-long vehicle would be lifted from the ground by two sleek solid rocket boosters and accelerate to 700 miles an hour within fifty seconds. After two minutes, the boosters would fall away and the shuttle's liquid-fueled main engines would complete the drive to orbit; the whole trip, from Florida launch pad to space, would take about nine minutes. At some 170 miles above Earth, the orbiter would cruise on its own momentum, using its on-board rocket maneuvering engines to adjust its path. It would travel at a steady 17,500 miles an hour—thirty times faster than a commercial jetliner—a speed that would carry it around the Earth in ninety minutes.

On its return, aided by corrective blasts from its maneuvering engines, the shuttle would hurtle into the atmosphere at twenty-five times the speed of sound, generating such friction that the insulation along its wing edges could reach 2,500 degrees Fahrenheit. Protected by 30,000-odd silica-fiber tiles, crew members inside would feel nothing of that outer blaze. Instead, using a series of S-turns, the pilot would gradually reduce the spacecraft's speed from a landing approach rate of 8,000 miles per hour to a moderate 200, coasting down to a one-and-a-half-mile roll on a runway. The landing would rely entirely on the pilot, for the shuttle was designed without air-breathing engines. Pilots call this kind of powerless, gliderlike descent a "dead-stick" landing. "Fighter pilots are crazy enough to enjoy the idea," said former astronaut Michael Collins. "And most astronauts are former fighter pilots."

The shuttle *Columbia* was the first to fly, making a practice run on April 12, 1981. The crew, veteran astronaut John Young and rookie Robert Crippen, stayed in orbit for two days, putting *Columbia* through its paces in nearly

1,000 small tests while circling the Earth some thirty-six times. That flight was followed by three other cautious trial voyages.

With the fifth launch, on November 11, 1982, NASA started putting the shuttle to work. The crew, Vance Brand, Bob Overmyer, Bill Lenoir, and Joseph Allen, dubbed themselves the "Ace Moving Company—Fast and Courteous Service." They deployed two communications satellites stored in the cargo bay and, as Allen said, "initiated a new era in which the business of space flight became business itself." A more efficient shuttle named *Challenger* joined *Columbia* in 1983, and the two carried several dozen satellites aloft in the mid-1980s; in 1985 alone the shuttles placed eleven satellites in orbit, on behalf of eight organizations from five countries.

The astronauts turned out to be far more than truckers. As well as deploying satellites, they carried out some delicate and risky space repairs and retrievals. In November 1984, NASA sent up a team to capture and bring down two crippled communications satellites belonging to Western Union and the Indonesian government. Allen, again a member of the crew, and mission specialist Dale Gardner strapped on motorized maneuvering units, floated out of the shuttle, and wrestled the satellites into the cargo bay, pretending they were lassoing cattle in a rodeo.

Like their colleagues aboard the Soviet space stations, the shuttle astronauts used the weightless environment as a scientific testing ground. Four missions between 1983 and 1985 carried Spacelab, a modular laboratory built by the European Space Agency. Tucked neatly in the shuttle's cargo, the lab housed over seventy experiments, from the creation of hormones to the growing of large, virtually flawless crystals in microgravity.

In 1986, however, tragedy brought all the congratulations to a halt with what should have been just another routine launch. The January 28 flight of the *Challenger* was a frightening reminder that even though space flight had begun to seem routine, it was in fact still a perilous enterprise.

**FIRE AND ICE**

The launch seemed flawless on that brilliantly cold January morning. The crew included Christa McAuliffe, the nation's first teacher in space, as well as commander Francis R. Scobee, pilot Michael Smith, and mission specialists Judith Resnick, Ellison Onizuka, Ronald McNair, and Gregory Jarvis. *Challenger* rose from the Florida coast in a liftoff so picture perfect that mission controllers in Houston at first missed the computer warning—a suddenly frozen screen—telling them that something had gone very wrong.

Seventy-three seconds after liftoff, the shuttle exploded into a cloud of burning debris, blasting apart when its fuel tanks caught on fire. All seven astronauts were killed. They may have had one fleeting, split-second warning of disaster: The last words of pilot Smith were a simple "uh-oh."

Investigators later discovered that burning gases had seeped out of one of the rocket boosters, escaping because O-ring seals in the rocket's aft joint had grown stiff with cold and so failed to seal properly. Richard Feynmann, a

Nobel laureate and a member of a presidential commission assigned to probe the issue, later dramatized the problem: Several minutes after he dunked a piece of rubber seal into ice water, the material hardened and became useless. The commission also found that at least some decision makers at NASA had known for years about problems with the O-rings but had not corrected them. On the day of the *Challenger* launch, engineers from Morton Thiokol, the manufacturer of the solid rocket boosters, had argued against liftoff in extreme cold because of just that risk. Their objections were overruled by managers at both Thiokol and NASA.

The *Challenger* disaster shocked the world. For NASA, it meant a painful period of self-blame and cautious regrouping that occupied almost three years. Yet to some, the incident was almost to be expected. Michael Collins, a member of the famous *Apollo 11* team, pointed out that the *Challenger* accident only emphasized what every space traveler already knew: Space flight is a high-risk undertaking. "If someone had suggested to me, in 1963 when I first became an astronaut, that for the next twenty-three years none of us would get killed riding a rocket, I would have said that person was a hopeless optimist and naive beyond words."

By September of 1988, the Americans had once again geared up for a gamble. The shuttle *Discovery* took off on September 29 and landed four days later in California, trailing a triumphant banner of white dust. Almost two months later, the new shuttle *Atlantis* also soared into orbit to deploy a defense satellite, marking a return to a regular schedule of orbiting flights.

On the heels of America's reentry into space, the Soviets unveiled their own space shuttle, *Buran.* On November 15, 1988, they flew the craft in a successful test with no humans aboard. Although it resembles U.S. shuttles, the snub-nosed vehicle lacks its own propulsion: All main engine power belongs to the huge Energia booster. Soviet planners would like the shuttle to serve as a ferry between Earth and *Mir 2*, a new station planned for 1994.

Shuttles and space stations are fast becoming permanent features of an ambitious international space program. By the end of 1989, orbiters are scheduled to deploy probes to Venus and Jupiter and to launch the long-awaited Hubble Space Telescope. A U.S. shuttle would also service a large space station to be built in the late 1990s by the United States, the European Space Agency, Canada, and Japan.

Yet even that station may be only a steppingstone to greater ventures. Cosmonaut Valeri Ryumin has already declared his willingness to spend a year in Earth orbit if it will be an intermediate stage in a voyage to Mars. Working from a reliable base in space, engineers could construct the sizable vehicles necessary to take travelers to that planet or even to the moons of Jupiter or Saturn. Expeditions from midrange space stations might also mine the rocky asteroid belt between Mars and Jupiter, or even look beyond the reaches of the Solar System itself. Thirty years after Yuri Gagarin rode singing into orbit, Earth's new-fledged space explorers might finally be ready to plan for the long flight to worlds circling another star.

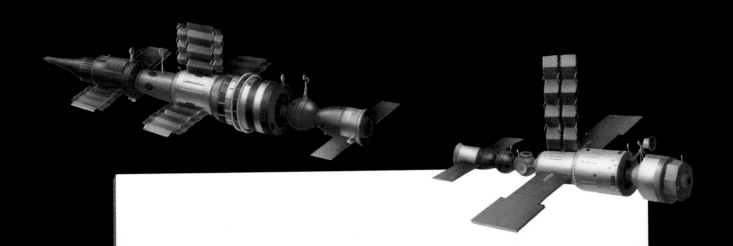

# LEARNING TO LIVE
# IN SPACE

Since 1971, dozens of cosmonaut teams aboard the Soviet space station Mir *(above, right)* and its Salyut predecessors *(above, left)* have endured cramped quarters, severely restricted activities, processed food, and the physical changes brought about by weightlessness while orbiting Earth for months at a time. The Soviet Union has undertaken these grueling flights in hopes that the cosmonauts' experiences will be the prelude to trips lasting not months but years, with other planets the destination.

One by one, problems that plagued earlier missions are being overcome. In the past, for instance, the noise of continually running ventilation fans disturbed the cosmonauts' sleep. By the early 1980s, insulation and design modifications had dropped station sound levels below forty decibels—quieter than a suburban night. The stations' lack of living space, which tends to trigger depression and irritability in cooped-up crew members, is being addressed with innovations such as private cabins. And Soviet physicians have developed strenuous physical regimens to combat the muscle atrophy, calcium depletion, and loss of appetite that occur in near zero gravity.

The stations still must be resupplied periodically with water, food, and other necessities from Earth, but if scientists and engineers can find ways to produce and recycle such essentials on board, a space station could become a self-sufficient community—and humankind will be free to venture into the unknown beyond the sustaining reach of the home world.

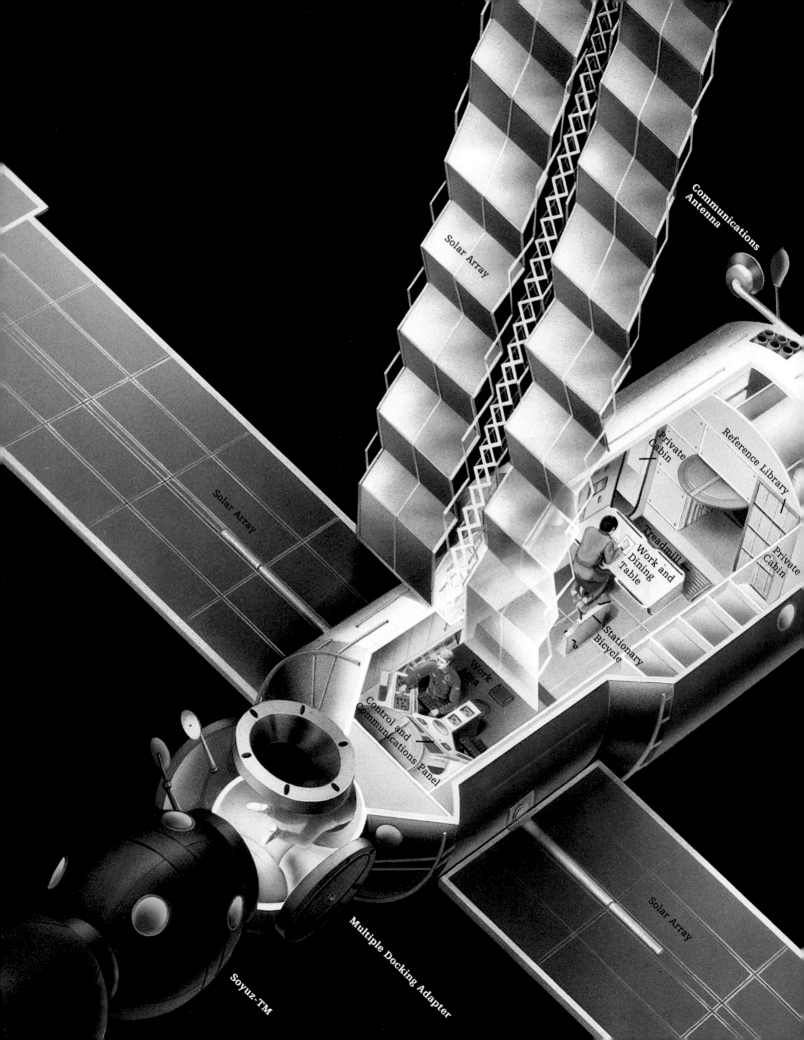

Communications
Antenna

Solar Array

Solar Array

Reference Library

Private
Cabin

Treadmill

Work and
Dining
Table

Private
Cabin

Stationary
Bicycle

Work
Area

Control and
Communications Panel

Solar Array

Multiple Docking Adapter

Soyuz-TM

Kvant
Module

Observation
Equipment

More than a thousand square feet of solar panels generate electrical power for equipment in the forward section, where cosmonauts do most of their work, and in the central area, where they eat, exercise, and relax. Various units may dock at one of six ports. Here, a Kvant research module has joined Mir at the single rear port, and a Soyuz-TM escape vehicle is at one of the five ports on the multiple docking adapter. A Soyuz-TM remains attached at all times in case of emergency.

# A COMPACT ORBITAL HOME

The state-of-the-art Soviet station Mir, launched in 1986, plays home to three resident cosmonauts and up to three guests in compact quarters totaling 5,300 cubic feet, including some elbowroom supplied by the so-called Kvant module ("quantum" in Russian), where astrophysical research projects are carried out. Other research modules are designed to attach to Mir from time to time at one of the station's several ports.

To create a sense of up and down in the strange world of near zero gravity, Mir's designers resorted to familiar visual cues. Chairs, tables, and most equipment hug the surface that is arbitrarily designated the floor, and white paint defines the ceiling. Color also gives character to other parts of the craft. Light green, believed to foster concentration, is used in the for-

ward work area, where the control and communications panels are located. Rearward of the work zone, a common room painted a cheerful yellow includes a dining area and pantry, a workbench, a treadmill, and a stationary bicycle.

Still further aft, soundproof curtains seal off two private cabins that are one of Mir's innovations; on earlier craft, cosmonauts merely attached sleeping bags to the walls. In a space about the size of a short telephone booth, each cabin contains a writing board, a sleeping bag, an intercom, and a porthole. Cosmonauts report that portholes are vital to morale: Few remedies for homesickness are more effective than the simple act of watching the cloud-swathed Earth, hundreds of miles below.

Yuri Romanenko readies his suit before taking a spacewalk to work on Mir's solar panels. Such outside excursions generally last no more than six hours.

Mission commander Vladimir Lyakhov moves cargo from a supply ship into Salyut 7's forward compartment. It takes a pair of cosmonauts as much as four days to unload the ferry.

# HOUSEKEEPING 225 MILES UP

With their lives dependent on the proper functioning of a host of complex systems, cosmonauts have become virtuosos of maintenance. The crew spends an hour or two every day at the forward control panels monitoring the space station's equipment. Whenever they replace or repair components, they use tools specially designed for zero gravity, including, for example, a hammer that does not recoil as an ordinary hammer would. Every three to six weeks, an unpiloted supply ship brings cargo that must be unloaded and stored—scientific instruments, food, water, letters from home, and fuel for the attitude-control thrusters. The cosmonauts then pack discarded equipment and other trash into the empty ship and launch it earthward to burn up on reentry into the atmosphere.

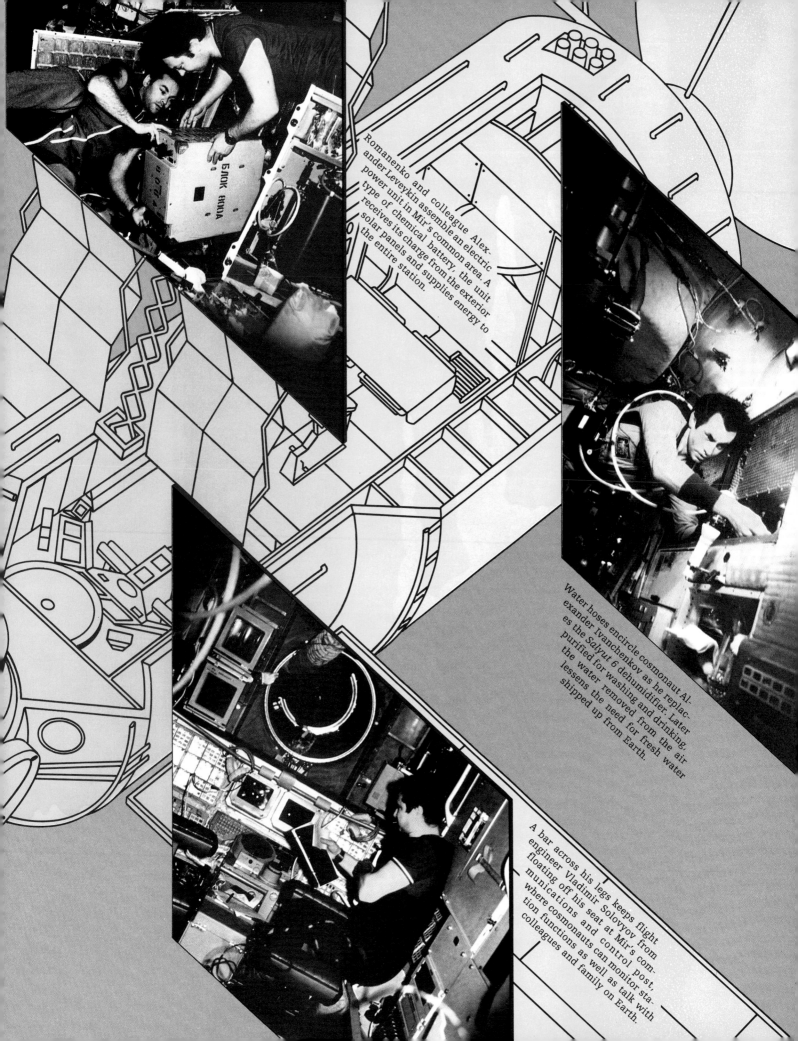

Romanenko and colleague Alexander Leveykin assemble an electric power unit in Mir's common area. A type of chemical battery, the unit receives its charge from the exterior solar panels and supplies energy to the entire station.

Water hoses encircle cosmonaut Alexander Ivanchenkov as he replaces the Salyut 6 dehumidifier. Later purified for washing and drinking, the water removed from the air lessens the need for fresh water shipped up from Earth.

A bar across his legs keeps flight engineer Vladimir Solovyov from floating off his seat at Mir's communications and control post, where cosmonauts can monitor station functions as well as talk with colleagues and family on Earth.

Romanenko's daily two-hour ride on the Mir exercise bike minimizes the loss of heart and skeletal muscle mass. The redistribution of body fluids in zero gravity causes the swollen face.

The appliance Salyut 7 cosmonaut Vladimir Vasyutin uses on Victor Savinykh simultaneously cuts his hair and vacuums up the clippings to keep them from drifting into station equipment. Their electric razors have a similar device.

Alexei Gubarev and Vladimir Remek dine standing up aboard Salyut 6. Most station food is canned or freeze-dried, but supply ships bring some fresh fruits and vegetables. With their meals, cosmonauts drink water, juice, tea, coffee, or vodka.

Anatoli Berezovoy and Valentin Lebedev set up the *Salyut 7* shower, a zippered cylinder that fills with a fine spray. Air sweeps the spray into a collector; the cosmonauts dry the booth to prevent mildew.

About to leave *Mir* after a six-month stay, flight engineer Alexander Alexandrov can risk getting upset if he loses a game of checkers to arriving crew member Musa Manarov. Grooved boards are used to keep game pieces in place.

Vladimir Kovalyonok moves a cluster of water receptacles aboard *Salyut 6*. Locked together, the receptacles form storage tanks. Cosmonauts sometimes drink from single receptacles using flexible plastic straws.

# THE FIT BODY AND MIND

Preserving the cosmonauts' physical and psychological well-being in zero gravity requires technological ingenuity and ceaseless effort. Because weightlessness impairs the immune system, for instance, cosmonauts take preventive action against microorganisms by sponging themselves and the station's walls with disinfectants every day. Eating is another challenge: To minimize floating crumbs, meals are coated with an edible film or come packaged in cans or tubes. In quarters where every inch counts, only a treadmill and stationary bicycle are necessarily sedate—for the crew's required exercise. Leisure pursuits are necessarily compact enough reading, talking to family on Earth, listening to music, watching video tapes. Ground controllers seldom allow competitive games, since conflict must be avoided as much as possible in such cramped surroundings.

127

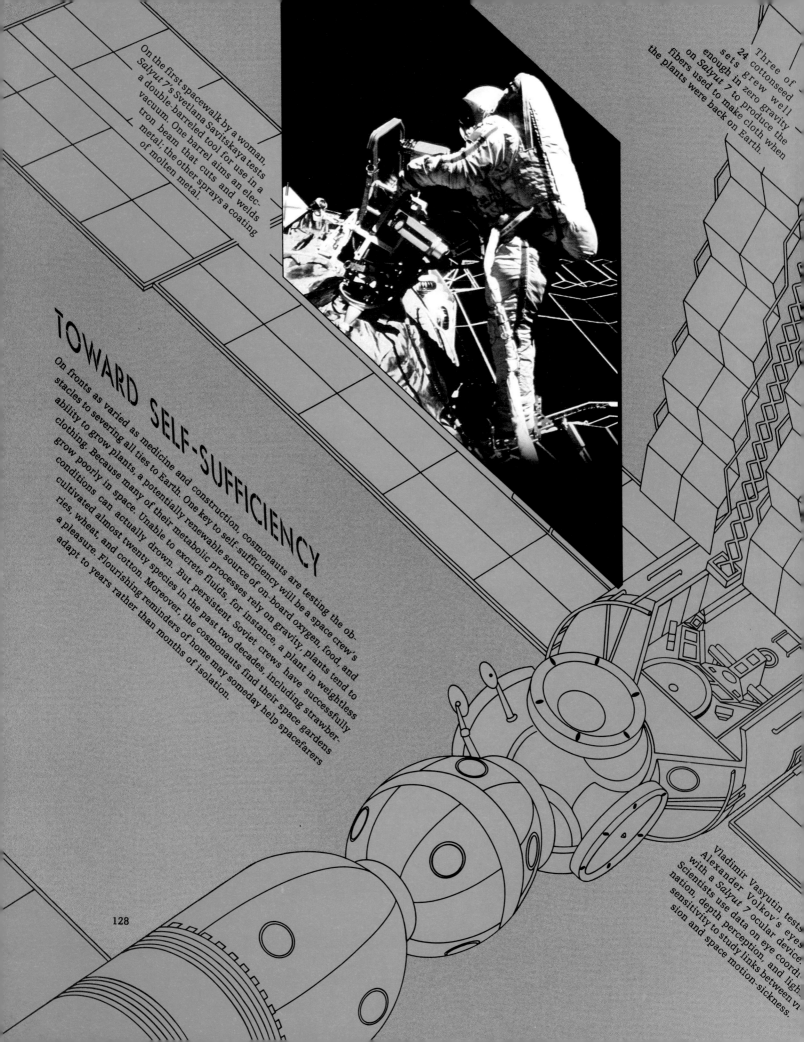

On the first spacewalk by a woman, *Salyut 7*'s Svetlana Savitskaya tests a double-barreled tool for use in a vacuum. One barrel aims an electron beam that cuts and welds metal; the other sprays a coating of molten metal.

Three of 24 cottonseed sets grew well enough in zero gravity on *Salyut 7* to produce the fibers used to make cloth when the plants were back on Earth.

# TOWARD SELF-SUFFICIENCY

On fronts as varied as medicine and construction, cosmonauts are testing the obstacles to severing all ties to Earth. One key to self-sufficiency will be a space crew's ability to grow plants, a potentially renewable source of on-board oxygen, food, and clothing. Because many of their metabolic processes rely on gravity, plants tend to grow poorly in space. Unable to excrete fluids, for instance, a plant in weightless conditions can actually drown. But persistent Soviet crews have successfully cultivated almost twenty species in the past two decades, including strawberries, wheat, and cotton. Moreover, the cosmonauts find their space gardens a pleasure. Flourishing reminders of home may someday help spacefarers adapt to years rather than months of isolation.

Vladimir Vasyutin tests Alexander Volkov's eyes with a *Salyut 7* ocular device. Scientists use data on eye coordination, depth perception, and light sensitivity to study links between vision and space motion-sickness.

In Mir's Kvant module, cosmonaut Alexander Leveykin adjusts a compact automated device for making antibiotics for livestock. The capacity to manufacture a sophisticated medical arsenal on board will be essential to long space journeys.

Salyut 6's Victor Savinykh checks orchids in a terrarium equipped to record their growth in time-lapse photos. The flowers faded prematurely; weightlessness probably disrupted the excretion of wastes.

129

"**I**f we die, we want people to accept it.
We are in a risky business, and we hope
that if anything happens to us it will
not delay the program. The conquest of
space is worth the risk of life."

Virgil "Gus" Grissom, *Apollo 1*, January 1967

**January 28, 1986.** One minute and
thirteen seconds after liftoff, the space
shuttle *Challenger* explodes, killing all
seven of the astronauts aboard.

**September 29, 1988.** Two years and eight months after the *Challenger* disaster, the space shuttle *Discovery* lifts off, returning American astronauts to space.

# GLOSSARY

**Acceleration:** a change in velocity. The term includes changes of direction and decreases as well as increases in speed.

**Accelerometer:** a device that senses changes in speed along an axis.

**Aeronautics:** the science of building and operating vehicles for flight.

**Aileron:** a hinged surface on the wing of an aircraft or spacecraft used to adjust the craft's angle of flight.

**Air lock:** a compartment separating areas of different environment, especially different air pressures, that is used for entry to and departure from a spacecraft.

**Altimeter:** a device that measures altitude above the surface of a planet or moon. Spacecraft altimeters work by timing the round trip of radio signals bounced off the surface.

**Apogee:** the point in the orbit of a satellite or other object when it is farthest from the Earth. *See* Perigee.

**Apolune:** the point in the orbit of a spacecraft when it is farthest from the Moon. *See* Perilune.

**Ascending node:** the point at which an orbiting object or spacecraft, traveling from south to north, crosses the plane of the equator. *See* Descending node.

**Astronaut:** a pilot or passenger on a space flight; also, a person in training for space flight. Soviet astronauts are called cosmonauts.

**Astronautics:** the science of operating a vehicle in space.

**Atmospheric pressure:** the weight of air on surfaces within Earth's atmosphere, about 14.7 pounds per square inch at sea level. Such pressure is also supplied artificially in spacecraft and spacesuits.

**Attitude:** a spacecraft's orientation with respect to its direction of motion.

**Autopilot:** a system or device that controls a vehicle's flight at a preset course and altitude.

**Azimuth:** an angular measure of an object's position along the horizon, described in degrees from west to east. As one of the coordinates expressing celestial location, it is sometimes used in tracking spacecraft. *See* Elevation.

**Ballistics:** the science of projectiles and their motion in flight.

**Black powder:** a mixture of saltpeter (potassium nitrate), sulfur, and charcoal, used in explosives and as an early propellant for rockets.

**Booster:** a rocket used to launch spacecraft.

**Bow shock wave:** the compressed wave that forms in front of a spacecraft or satellite as it moves rapidly through Earth's atmosphere; more generally, any such wave that forms between an object and a fluid medium.

**Burn:** combustion action in rockets. Propulsion in space is achieved through a sequence of burns.

**C-band:** a radio frequency of 5 gigahertz used for spacecraft communications on Mercury and Gemini missions.

**Celestial sphere:** the apparent sphere of sky that surrounds the Earth; used as a convention for specifying the location of a celestial object.

**Combustion:** a chemical reaction between two or more substances that releases heat, light, and gases.

**Command module:** the section of an Apollo spacecraft that carries the crew and communications equipment and is used as the reentry vehicle.

**Countdown:** a checking procedure used before rocket and spacecraft launches. Standard time units are marked off by counting backward toward zero; during each unit the operation of specific items or systems is checked. Launch takes place at zero.

**Corona:** the Sun's outer layer. The corona's changing appearance reflects changing solar activity.

**Deceleration:** negative acceleration, slowing.

**Declination:** One of the coordinates, measured in degrees, used to designate the location of a fixed object, such as a star, on the celestial sphere. Declination is a north-south value similar to latitude on Earth. *See* Right ascension.

**Descending node:** the point at which an orbiting object or spacecraft, moving from north to south, crosses the plane of the equator.

**Descent engine:** the rocket used to power a spacecraft as it makes a controlled landing on the surface of a planet or moon.

**Doppler effect:** a phenomenon in which waves appear to compress as their source approaches the observer or stretch out as the source recedes from the observer.

**Eccentricity:** the amount of separation between the two foci of an ellipse and, hence, the degree to which an elliptical orbit deviates from a circular shape.

**Elevation:** an angular measure of the height of an object above the horizon; with azimuth, one of the coordinates defining celestial location and sometimes used in tracking spacecraft.

**Engine:** in spacecraft, a rocket or thruster that burns liquid propellants and can be throttled to adjust thrust.

**Escape tower:** a rocket-powered framework designed to separate spacecraft modules from their booster rockets in case of accident. Escape towers are mounted atop the spacecraft and jettisoned after launch.

**Extravehicular activity (EVA):** action performed by an astronaut or cosmonaut outside a vehicle in space; a spacewalk.

**Frequency:** the number of oscillations per second of an electromagnetic (or other) wave.

**Fuel:** a substance that when combined with an oxidizer burns to produce thrust in rockets.

**Geosynchronous:** describing the orbit of a spacecraft or satellite that completes a circle every twenty-four hours, the same time Earth requires to make one rotation; thus the object remains above one location on the ground. Geosynchronous orbits are established over the equator at an altitude of approximately 22,300 miles.

**Gimbal:** a device designed to hold an object or an instrument such as a spacecraft's inertial measurement unit at a constant angle within a moving framework.

**Grain:** the rubberlike mass of chemical propellant that provides propulsion in solid fuel rockets. The shape of the grain determines the rate and pattern of burn and thus controls thrust.

**Gravity:** the force responsible for the mutual attraction of separate masses. *See* Microgravity, Zero gravity.

**Gyroscope:** a spinning, wheel-like device that resists any force that tries to tilt its axis. Gyroscopes are used for stabilizing the attitude of rockets and spacecraft in motion.

**Inertial guidance:** the in-flight guidance of a rocket or spacecraft by automatic devices following a programmed flight path.

**Inertial measurement unit (IMU):** an on-board instrument system that measures the attitude of a spacecraft. It includes accelerometers and gyroscopes.

**Injection angle:** the angle at which a spacecraft's return trajectory intersects the Earth's atmosphere.

**Inverse-square law:** the mathematical description of how the strength of some forces, including gravity, changes in inverse proportion to the square of the distance from the source.

**Ionosphere:** an atmospheric layer dominated by charged, or ionized, atoms that extends from about 38 to 400 miles above the Earth's surface.

**K-band:** a communications frequency of 15 gigahertz used for high-speed data transmission on shuttle flights.

**Kinetic energy:** an object's energy of motion; for example, the force of a falling body.

**Lunar module:** the craft used by Apollo missions for Moon landings. The lunar module consisted of a descent stage, used to land on the Moon and as a platform for liftoff, and an ascent stage, used as crew quarters and for returning to the orbiting command module.

**Magnetometer:** a device for measuring the strength and direction of a magnetic field.

**Manned maneuvering unit (MMU):** a portable jet-pack device used by astronauts to propel themselves through space independent of a spacecraft.

**Microgravity:** an environment of very weak gravitational forces, such as those within an orbiting spacecraft. Microgravity conditions in space stations may allow experiments or manufacturing processes that are not possible on Earth.

**Motor:** in spacecraft, a rocket that burns solid propellants. *See* Solid rocket booster.

**Orbit:** the path of an object revolving around another object; also, to progress along such a periodic path.

**Oxidizer:** an agent that releases oxygen for combination with another substance, creating combustion and gas for propulsion.

**Parking orbit:** a temporary orbit for a spacecraft or satellite, established for rendezvous with other spacecraft, for awaiting advantageous trajectories, or as part of modular missions such as the Apollo Moon landings.

**Payload:** revenue-producing or useful cargo carried by a spacecraft; also, anything carried in a rocket or spacecraft that is not part of the structure, propellant, or guidance systems.

**Perigee:** the point in the orbit of a spacecraft or satellite when it is closest to Earth. *See* Apogee.

**Perilune:** the point in the orbit of a spacecraft when it is closest to the Moon. *See* Apolune.

**Pitch:** the movement of a craft about its lateral axis; the nose pitches up or down about this axis.

**Plasma:** a gaslike association of ionized particles that responds collectively to electric and magnetic fields.

**Plutonium-238:** a form of the radioactive element plutonium, characterized by high energy emissions.

**Polymer:** a compound used as a binder for rocket propellant systems; more generally, a compound consisting of repeating structural units.

**Propellant:** a chemical or chemical mixture burned to create the thrust for a rocket or spacecraft.

**Radiation:** energy in the form of electromagnetic waves or particles.

**Radio:** the least energetic form of electromagnetic radiation, having the lowest frequency and the longest wavelength.

**Reentry:** the descent into Earth's atmosphere from space.

**Retrorocket:** a rocket that produces thrust in a direction opposite to the direction of motion in order to slow a larger rocket or satellite.

**Right ascension:** with declination, one of the coordinates used to designate the location of a fixed object, such as a star, on the celestial sphere. Right ascension is measured in hours, minutes, and seconds and is similar to longitude on Earth.

**Rocket:** a missile or vehicle propelled by the combustion of a fuel and a contained oxygen supply. The forward thrust of a rocket results when exhaust products are ejected from the tail.

**Roll:** motion about the longitudinal axis of a flying body.

**Satellite:** any body, natural or artificial, in orbit around a planet. The term is used most often to describe moons and spacecraft.

**S-band:** a radio frequency of 2 gigahertz used for communicating with piloted space missions.

**Seismometer:** a device for measuring movements of the ground.

**Service module:** the self-contained section of the Apollo spacecraft that housed thrusters, air and water, and the electric power supply.

**Sextant:** an instrument that measures angular distances from fixed celestial objects, sometimes used in spacecraft as a navigational aid.

**Simulator:** a device that mimics the operational conditions of equipment or vehicles.

**Solar panel:** an array of light-sensitive cells attached to a spacecraft and used to generate electrical power for the vehicle in space.

**Solar wind:** a current of charged particles that streams outward from the Sun.

**Solid rocket booster:** a rocket, powered by solid propellants, used to launch spacecraft into orbit.

**Space:** the universe beyond Earth's atmosphere. The boundary at which the atmosphere ends and space begins is not sharp but starts at approximately 100 miles above Earth's surface.

**Spacecraft:** a piloted or unpiloted vehicle designed for travel in space.

**Space station:** an orbiting spacecraft designed to support human activity for an extended time.

**Stage:** an independently powered section of a rocket or spacecraft, often combined with others to form multistage vehicles.

**Subatomic particles:** fundamental components of matter such as electrons or protons.

**Sublimator:** an exposed metal plate, located on the outside of a spacesuit, that functions as a cooling coil to control suit temperatures.

**Telemetry:** data, usually measurements, transmitted from a remote sensor to a recording receiver.

**Thermal energy:** energy in the form of heat.

**Throttle:** to decrease the supply of propellant to an engine, reducing thrust. Liquid propellant rocket engines can be throttled; solid rocket motors cannot.

**Thrust:** the force that propels a rocket or spacecraft. Thrust is generated by a high-speed jet of gases discharging through a nozzle.

**Thrusters:** rocket engines used for maneuvering spacecraft in space.

**Tracking:** the science of monitoring satellite locations by means of radio antennas at ground stations or by using other satellite systems in space.

**Trajectory:** the curve traced by an object moving through space. A closed trajectory is an orbit.

**Transponder:** a device that transmits a response signal automatically when activated by an incoming signal.

**Ultrahigh frequency (UHF):** short radio waves used for communicating with spacecraft.

**Ultraviolet:** a band of electromagnetic radiation with a higher frequency and shorter wavelength than visible blue light. Ultraviolet astronomy is generally performed in space, since Earth's atmosphere absorbs most ultraviolet radiation.

**Van Allen belts:** two doughnut-shaped zones of radiation about the Earth, concentrated at altitudes of 3,000 and 10,000 miles. The belts contain charged particles generated by solar flares and trapped by the Earth's magnetic field.

**Wavelength:** the distance from crest to crest, or trough to trough, of an electromagnetic or other wave. Wavelengths are related to frequency: The longer the wavelength, the lower the frequency.

**X-rays:** a band of electromagnetic radiation intermediate in wavelength between ultraviolet radiation and gamma rays. Because x-rays are absorbed by the atmosphere, x-ray astronomy is performed in space.

**Yaw:** the movement of a craft about its vertical axis; the nose and tail move from side to side.

**Zero gravity:** a condition in which gravity appears to be absent. Zero gravity occurs when gravitational forces are balanced by the acceleration of a body in orbit or free fall.

# BIBLIOGRAPHY

**Books**

Adelman, Saul J., and Benjamin Adelman, *Bound for the Stars.* Englewood Cliffs, N.J.: Prentice-Hall, 1981.

Allen, Joseph P., and Russell Martin, *Entering Space.* New York: Stewart, Tabori & Chang, 1985.

Allen, Oliver E., and the Editors of Time-Life Books, *Atmosphere* (Planet Earth series). Alexandria, Va.: Time-Life Books, 1983.

Arno, Roger, *The Story of Space & Rockets.* Santa Barbara, Calif.: Bellerophone Books, 1986.

Battin, Richard H., *Astronautica Guidance.* New York: McGraw-Hill, 1964.

Bond, Aleck C., and Maxime A. Faget, *Technologies of Manned Space Systems.* New York: Gordon and Breach, 1965.

Bond, Peter, *Heroes in Space.* New York: Basil Blackwell, 1987.

Borman, Frank, and Robert J. Serling, *Countdown: An Autobiography.* New York: Silver Arrow Books, 1988.

Burchett, Wilfred, and Anthony Purdy, *Cosmonaut Yuri Gagarin: First Man in Space.* London: Anthony Gibbs & Phillips, 1961.

Carpenter, M. Scott, et al., *We Seven.* New York: Simon and Schuster, 1962.

Clark, Phillip, *The Soviet Manned Space Program.* New York: Salamander Books, 1988.

Collins, Michael:
*Carrying the Fire: An Astronaut's Journeys.* New York: Farrar, Straus and Giroux, 1974.
*Liftoff.* New York: Grove Press, 1988.

Corliss, William R.:
*Spacecraft Tracking.* Washington, D.C.: NASA, 1968.
*Space Probes and Planetary Exploration.* Princeton, N.J.: D. Van Nostrand, 1965.

Cortright, Edgar M., *Apollo Expeditions to the Moon.* Washington, D.C.: NASA, 1975.

Cowart, E. G., *Lunar Rover Vehicle.* Huntsville, Ala.: Boeing, 1975.

Draper, C. S., et al., *Space Navigation Guidance and Control.* Maidenhead, England: Technivision Ltd., 1966.

Emme, Eugene M., ed., *The History of Rocket Technology.* Detroit: Wayne State University Press, 1964.

Ezell, Edward Clinton, and Linda Neuman Ezell, *The Partnership: A History of the Apollo-Soyuz Test Project.* Washington, D.C.: NASA, 1978.

Froehlich, Walter, *Apollo Soyuz.* Washington, D.C.: NASA, 1976.

Furniss, Tim, *Manned Spaceflight Log.* London: Jane's, 1983.

Gatland, Kenneth, *The Illustrated Encyclopedia of Space Technology.* New York: Salamander Books, 1981.

Glasstone, Samuel, *Sourcebook on the Space Sciences.* Princeton, N.J.: D. Van Nostrand, 1965.

Hartmann, William K., *Astronomy: The Cosmic Journey.* Belmont, Calif.: Wadsworth, 1987.

Hooper, Gordon R., *The Soviet Cosmonaut Team.* Woodbridge, Suffolk, England: GHR Publications, 1986.

Hunter, Maxwell W., II, *Thrust into Space.* New York: Holt, Rinehart and Winston, 1966.

Hurt, Harry, III, *For All Mankind.* New York: Atlantic Monthly Press, 1988.

Huzel, Dieter K., *Peenemünde to Canaveral.* Englewood Cliffs, N.J.: Prentice-Hall, 1962.

Kerrod, Robin, *The Illustrated History of NASA.* New York: Gallery Books, 1986.

Krieger, F. J., *Behind the Sputniks: A Survey of Soviet Space Science.* Washington, D.C.: Public Affairs Press, 1958.

Lehman, Milton, *This High Man: The Life of Robert H. Goddard.* New York: Farrar, Straus, 1963.

Lewis, Richard S.:
*Appointment on the Moon.* New York: Viking Press, 1968.
*The Voyages of Apollo: The Exploration of the Moon.* New York: Quadrangle/The New York Times Book Company, 1974.

Ley, Willy:
*Events in Space.* New York: David McKay, 1969.
*Rockets, Missiles, and Men in Space.* New York: Viking Press, 1968.
*Satellites, Rockets and Outer Space.* New York: New American Library, 1962.

*Life in Space,* by the Editors of Time-Life Books. Alexandria, Va.: Time-Life Books, 1983.

McDougall, Walter A., . . .*The Heavens and the Earth.* New York: Basic Books, 1985.

*The McGraw-Hill Encyclopedia of Space.* New York: McGraw-Hill, 1968.

Mallan, Lloyd, *Suiting Up for Space.* New York: John Day, 1971.

Moore, Patrick, *Space.* Garden City, N.Y.: Natural History Press, 1968.

Newell, Homer E., *Express to the Stars.* New York: McGraw-Hill, 1961.

Nicks, Oran W., *Far Travelers.* Washington, D.C.: NASA, 1985.

Nicogossian, Arnauld E., and James F. Parker, Jr., *Space Physiology and Medicine.* Washington, D.C.: NASA, 1982.

Oberg, James E.:
*The New Race for Space.* Harrisburg, Pa.: Stackpole Books, 1984.
*Red Star in Orbit.* New York: Random House, 1981.

Ordway, Frederick I., III, and Mitchell R. Sharpe, *The Rocket Team.* New York: Thomas Y. Crowell, 1979.

Osman, Tony, *Space History.* New York: St. Martin's Press, 1983.

Park, Robert A., and Thomas Magness, *Interplanetary Navigation.* New York: Holt, Rinehart and Winston, 1964.

Pellegrino, Charles R., and Joshua Stoff, *Chariots for Apollo.* New York: Atheneum, 1985.

Riley, Francis E., and J. Douglas Sailor, *Space Systems Engineering.* New York: McGraw-Hill, 1962.

Shankle, Ralph O., *The Twins of Space: The Story of the Gemini Program.* Philadelphia: J. B. Lippincott, 1964.

Sheldon, Charles S., II, *Review of the Soviet Space Program.* New York: McGraw-Hill, 1968.

Shepherd, Dennis G., *Aerospace Propulsion.* New York: American Elsevier, 1972.

Slukhai, I. A., *Russian Rocketry: A Historical Survey.* Jerusalem: Israel Program for Scientific Translations, 1968.

Stearns, Edward V. B., *Navigation and Guidance in Space.* Englewood Cliffs, N.J.: Prentice-Hall, 1963.

Stoiko, Michael, *Soviet Rocketry.* New York: Holt, Rinehart and Winston, 1970.

Sutton, George P., and Donald M. Ross, *Rocket Propulsion Elements.* New York: John Wiley & Sons, 1976.

Swenson, Loyd S., Jr., James M. Grimwood, and Charles C. Alexander, *This New Ocean.* Washington, D.C.: NASA, 1966.

Thomas, Shirley, *Satellite Tracking Facilities.* New York: Holt, Rinehart and Winston, 1963.

Thruelsen, Richard, *The Grumman Story.* New York: Praeger, 1976.

Titov, Gherman, and Martin Caidin, *I Am Eagle!* Indianapolis: Bobbs-Merrill, 1962.

Turnill, Reginald, *Jane's Spaceflight Directory.* London: Jane's, 1984.

Von Braun, Wernher, *Space Frontier.* New York: Holt, Rinehart and Winston, 1971.

Von Braun, Wernher, and Frederick I. Ordway III:
*History of Rocketry & Space Travel.* New York: Thomas Y. Crowell, 1975.
*The Rockets' Red Glare.* Garden City, N.Y.: Anchor Press, 1976.

Walters, Helen B.:
*Hermann Oberth: Father of Space Travel.* New York: Macmillan, 1962.
*Wernher Von Braun: Rocket Engineer.* New York: Macmillan, 1965.

White, Frank, *The Overview Effect.* Boston: Houghton Mifflin, 1987.

Winter, Frank H., *Prelude to the Space Age: The Rocket Societies, 1924-1940.* Washington, D.C.: Smithsonian Institution Press, 1983.

Wolff, Edward A., *Spacecraft Technology.* Washington, D.C.: Spartan Books, 1962.

Young, Hugo, Bryan Silcock, and Peter Dunn, *Journey to Tranquility.* Garden City, N.Y.: Doubleday, 1969.

Zeilik, Michael, and Elske v.P. Smith, *Introductory Astronomy and Astrophysics.* Philadelphia: Saunders College Publishing, 1987.

**Periodicals**

"Apollo 11 Lunar Landing Mission Profile." *Aviation Week & Space Technology,* July 7, 1969.

"Boeing Wins Lunar Rover Award." *Aviation Week & Space Technology,* November 3, 1969.

Bond, Constance, "The Ancestors—and Lost Cousins—of the Space Suit." *Smithsonian,* April 1983.

Canby, Thomas Y., "Soviets in Space: Are They Ahead?" *National Geographic,* October 1986.

Cernan, Eugene, Ronald Evans, and Harrison Schmitt (crew of *Apollo 17),* "The Final Flight." *National Geographic,* September 1973.

Collins, Michael, " 'The Stars Are Everywhere, Even Below Me!' " *Smithsonian,* May 1988.

Cowley, Geoffrey, et al., " 'Liftoff, Liftoff.' " *Newsweek,* October 10, 1988.

"Dual-Mode Lunar Rover Usefulness Cited." *Aviation Week & Space Technology,* January 6, 1969.

Gore, Rick, "When the Space Shuttle Finally Flies." *National Geographic,* March 1981.

Hall, Alice J., "Apollo 14: The Climb Up Cone Crater." *National Geographic,* July 1971.

Hill, Arthur, "Flying Taxis for Moon Explorers." *Space World,* July 1968.

Jaroff, Leon, "The Magic Is Back!" *Time,* October 10, 1988.

"JPL Outlines Requirements for Lightweight Roving Lunar Vehicles." *Aviation Week & Space Technology,* July 1, 1963.

Klass, Philip J.:
"Apollo Optical-Inertial Guidance Detailed." *Aviation Week & Space Technology,* September 30, 1963.
"Inertial Navigation: Out of the Laboratory." *Aviation Week,* January 2, 1956.

Magnuson, Ed, "They Slipped the Surly Bonds of Earth to Touch the Face of God." *Time,* February 10, 1986.

Marbach, William D., et al., "What Went Wrong?" *Newsweek,* February 10, 1986.

Melbourne, William G., "Navigation between the Planets." *Scientific American,* June 1976.

"Modern Mule for the Lunar Prospector." *Space World,* January 1969.

"The 'Mooncopter' Lunar Exploration Aid." *Space World,* October 1968.

"Moon Wheels for the Astronauts." *Business Week,* August 2, 1969.

Pokrovskiy, A., "Report from 'Mir' Orbital Station Trainer." *Pravda,* February 21, 1986.

Richey, B. J., "Moon Vehicle Is Necessity for Real Lunar Exploration." *Huntsville Times,* March 9, 1969.

Shifrin, Carole A., "NASA to Evaluate Two Suit Designs for Space Station." *Aviation Week & Space Technology,* January 11, 1988.

"Soviet Mir Mockup Used for Training." *Aviation Week & Space Technology,* October 6, 1986.

Strickland, Z., "Bendix, Boeing Selected to Bid on Lunar Rover." *Aviation Week & Space Technology,* October 6, 1969.

Taylor, Richard L. S., "Manned Spaceflight: The Human Barrier." *Space,* January/February 1988.

Thompson, Richard, "Surging Ahead." *Time,* October 5, 1987.

"USSR Plans New Mir Crew Launch, Docking of Building-Block Module." *Aviation Week & Space Technology,* October 20, 1986.

Von Braun, Wernher, "Dr. von Braun: Travel on the Moon." *Popular Science,* February 1964.

Vosburgh, Frederick G., "Flying in the 'Blowtorch' Era." *National Geographic,* September 1950.

Walker, T. W., "The Development of the Pressure Suit for High Altitude Flying." *Project Engineer,* May 1956.

Weiss, E. H., "Tracking Earth Satellites." *Byte,* July 1985.

Wellborn, Stanley N., et al., "Out of Challenger's Ashes Full Speed Ahead." *U.S. News & World Report,* February 10, 1986.

Wetmore, Warren C., "Trajectory, Timing of Apollo 11 Revised to Permit Goldstone Antenna to Cover Landing." *Aviation Week & Space Technology,* July 14, 1969.

Whitehouse, David, "Buran Swirls In." *Space,* January/February 1989.

Wilson, K. T., "Spacesuit Development: The American Experience." *Journal of the British Interplanetary Society,* 1985, vol. 38, pages 51-60.

## Other Publications

Alexander, James D., and Robert W. Becker, "Evolution of the Rendezvous-Maneuver Plan for Lunar-Landing Missions." Apollo Experience Report. Houston, Tex.: NASA, Lyndon B. Johnson Space Center, April 25, 1973.

"Apollo News Reference Handbook." Bethpage, N.Y.: Grumman Aircraft Engineering Corporation, 1969.

"Apollo Operations Handbook: Subsystems Data." Mission LM Report. Washington, D.C.: NASA, February 1, 1970.

"Apollo Program Summary Report." Houston, Tex.: NASA, Lyndon B. Johnson Space Center, April 1975.

"Apollo 17: Guidance and Navigation Briefing." Santa Barbara, Calif.: Delco Electronics, December 1972.

Bluth, B. J., and Martha Helppie, "Soviet Space Stations as Analogs" (2d ed.). Washington, D.C.: NASA, August 1986.

Committee on Commerce, Science, and Transportation, U.S. Senate, "Soviet Space Programs: 1981-87." Committee Print. Washington, D.C.: Government Printing Office, May 1988.

Corliss, William R., "History of the Goddard Networks." Greenbelt, Md.: Goddard Space Flight Center, November 1, 1969.

Emme, Eugene M., "Robert H. Goddard: World Rocket Pioneer." NASA Historical Report No. 1. Washington, D.C.: NASA, July 1960.

Feoktistov, Konstantin, "The Courage of the 'Pamiry.'" Soviet Spaceflight Report. Starwise Publications, 1985.

Froehlich, Walter, "The Tracking & Data Relay Satellite System." Washington, D.C.: NASA, no date.

Hooke, Lydia Razran, et al., eds., "USSR Space Life Sciences Digest." Issue 16, NASA Contractor Report 3922(19). Washington, D.C.: NASA, 1988.

"Introduction to Gyroscopes." Inertial Instruments series. Santa Barbara, Calif.: Delco Systems Operations, 1987.

Johnson, Nicholas L., "The Soviet Year in Space: 1986." Colorado Springs: Teledyne Brown Engineering, no date.

Kelly, Thomas J., "Design Features of the Project Apollo Lunar Module." Presented at the Annual Meeting and Technical Display, American Institute of Aeronautics and Astronautics, New York, May 12-14, 1981.

Kozloski, Lillian D., "Suiting Up for Space: U.S. Space Suits of the National Air and Space Museum." NASM Artifact series. Washington, D.C.: NASM, in press.

Lauzon, Shelley M., "Robert Hutchings Goddard Memorial Lecture." Worcester, Mass.: Clark University, October 19, 1978.

Lowman, Paul D., Jr., "Lunar Bases and Post-Apollo Lunar Exploration: An Annotated Bibliography of Federally-Funded American Studies, 1960-82." Greenbelt, Md.: Goddard Space Flight Center, October 1984.

"The Lunar Mission Analysis Branch." In "The Apollo 11 Adventure," Marshall Space Flight Center Internal Note 70-FM-20, February 5, 1970.

"Manned Maneuvering Unit." Denver, Colo.: Martin Marietta Aerospace, March 1985.

"Mobile and Fixed Base Lunar Simulators." MOLAB Fact Sheet. New York: Grumman Aircraft, 1965.

"Salyut: Soviet Steps toward Permanent Human Presence in Space." Technical Memorandum. Washington, D.C.: U.S. Congress, Office of Technology Assessment, December 1983.

Smith, Marcia S., "Space Activities of the United States, Soviet Union and Other Launching Countries: 1957-1987." CRS Report for Congress. Washington, D.C.: Library of Congress, February 29, 1988.

"Space Navigation Handbook." Greenbelt, Md.: Goddard Space Flight Center, 1964.

"Space Shuttle: Space Suit/Life Support System." Windsor Locks, Conn.: United Technologies, Hamilton Standard, no date.

"Systems and Facilities." Vol. 1 of "National Space Transportation System Reference." Washington, D.C.: NASA, June 1988.

"U.S. and Soviet Manned Spaceflights, 1961-1986." In "NASA Aeronautics and Space Report to the President: 1986 Activities." Washington, D.C.: NASA, no date.

"U.S.-Soviet Cooperation in Space." Technical Memorandum. Washington, D.C.: U.S. Congress, Office of Technology Assessment, July 1985.

Wilde, Richard:
  "EMU—A Human Spacecraft." Presented at the Fourteenth International Symposium on Space Technology and Science, Tokyo, 1984.
  "Extravehicular Mobility Unit Description: Block II." Windsor Locks, Conn.: United Technologies, Hamilton Standard, March 1, 1984.

# INDEX

*Numerals in italics indicate an illustration of the subject mentioned.*

# ACKNOWLEDGMENTS

The editors of *Outbound* wish to thank the following individuals and organizations for their contributions to this volume: Gerald W. Bursek, Delco Electronics, Goleta, Calif.; Gerald Condon, Lyndon B. Johnson Space Center, Houston, Tex.; Georges Delaleau, Gilbert Deloizy, Musée de l'Air et de l'Espace, Le Bourget, France; Fred Haise, Grumman Space Station Program, Support Division, Reston, Va.; Martin Jenness, Lyndon B. Johnson Space Center, Houston, Tex.; C. C. Johnson, Space Industries, Webster, Tex.; Malcolm Jones, Lyndon B. Johnson Space Center, Houston, Tex.; Thomas J. Kelly, Grumman Corporation, Bethpage, N.Y.; James W. Kennedy, National Aeronautics and Space Administration, Washington, D.C.; Edgar Lineberry, Jr.; George Miller, Goddard Space Flight Center, Greenbelt, Md.; Andrew Patnesky, Lyndon B. Johnson Space Center, Houston, Tex.; Rocco Petrone, Rockwell International Corp., Seal Beach, Calif.; Lee Saegesser, National Aeronautics and Space Administration, Washington, D.C.; Emil Schiesser, Lyndon B. Johnson Space Center, Houston, Tex.; George Skurla, Grumman Corporation, Bethpage, N.Y.; Marcia Smith, Congressional Research Service, Washington, D.C.; Joseph Thibodeau, Lyndon B. Johnson Space Center, Houston, Tex.; Howard William Tindall, Martin Marietta, Washington, D.C.; Douglas K. Ward, Lyndon B. Johnson Space Center, Houston, Tex.; Paul F. Wetzel, National Aeronautics and Space Administration, Washington, D.C.

# PICTURE CREDITS

*The sources for the illustrations in this book are listed below. Credits from left to right are separated by semicolons; credits from top to bottom are separated by dashes.*

Cover: Art by Damon Hertig. Front and back endpapers: Art by John Drummond. 2, 3: NASA, no. SL3-119-2255. 4, 5: NASA, no. AS12-51-7507. 6, 7: NASA, no. AS11-40-5872. 8, 9: NASA, no. SL3-22-871. 14, 15: NASA. 16: Initial cap, detail from pages 14, 15. 18, 19: Science Museum, London/courtesy Major C. Congreve; The Mansell Collection, London; Smithsonian Institution, no. A4317E; Smithsonian Institution, no. 76-17287; Smithsonian Institution, no. A-3966C; courtesy Historical Division, Department of the Army, U.S. Army Missile Command; Smithsonian Institution, no. 76-7559; Smithsonian Institution, no. A-2458; Smithsonian Institution, no. 84-10327; Smithsonian Institution, no. 73-7133. Art by Stephen Wagner. 20, 21: Smithsonian Institution, no. 84-8949; Smithsonian Institution, no. A4558-C; Smithsonian Institution, no. A4561; U.S. Air Force Flight Test Center History Office (2). Art by Stephen Wagner. 25-29: Art by Stephen Wagner. 32, 33: from *Pokorenie Komosa,* USSR. 36-41: Art by Damon Hertig. 42, 43: NASA. 44: Initial cap, detail from pages 42, 43. 48, 49: Sovfoto; Henry Walker for *Life;* Sovfoto; NASA, no. MR3-45/61-MR3-107; NASA. Art by Stephen Wagner. 50, 51: TASS; Sovfoto; Novosti; Donald C. Uhrbrock for *Life;* NASA, no. 65-HC-361; NASA, no. C-692-GT7/74595; NASA, no. GTAS/75308. Art by Stephen Wagner. 56-59: Art by Stephen Bauer. 64-73: Art by Jeffrey Oh. 74, 75: NASA, no. AS17-152-23274. 76: Initial cap, detail from pages 74, 75. 79-82: Art by Stephen Bauer. 85: NASA, no. 66-HC-1827; NASA, no. 68-HC-225. Art by Stephen Wagner. 86, 87: NASA, no. 68-HC-730; NASA, no. 69-HC-164; NASA, no. 69-HC-541; NASA, no. 69-HC-810; NASA, no. 69-HC-808; NASA no. 69-HC-1236. Art by Stephen Wagner. 88, 89: NASA, no. 70-HC-323—NASA, no. 70-HC-990; NASA, no. 71-HC-85; NASA, no. 72-HC-149; NASA, no. 72-HC-907. Art by Stephen Wagner. 92: Grumman Corp. 93: NASA, no. MSFC-67-RDO-7451. 94, 95: NASA, no. 68-H-283/68-HC-165—Grumman Corp.; NASA, no. 71-HC-1143. 98-103: Art by Stephen Wagner. 104, 105: NASA, no. SL2-7-667. 106: Initial cap, detail from pages 104, 105. 108, 109: From *The National Air and Space Museum,* by C. D. B. Bryan, with photographs by Michael Freeman, Robert Golden, and Dennis Rolfe, published by Harry N. Abrams, Inc., New York, 1979. Background art by Stephen Wagner. 110, 111: National Air and Space Museum, courtesy Smithsonian Institution—Bettmann Archive; Smithsonian Institution, no. A43061-B; Musée de l'Air et de l'Espace, Le Bourget, France; Courtesy USAF, from *Suiting Up for Space: The Evolution of the Space Suit,* by Lloyd Mallan, published by The John Day Co., New York, 1971 (5). 112, 113: Larry Sherer, courtesy USAF, from *Suiting Up for Space: The Evolution of the Space Suit,* by Lloyd Mallan, published by The John Day Co., New York, 1971 (2); International Latex Corp.; Larry Sherer, courtesy USAF, from *Suiting Up for Space: The Evolution of the Space Suit,* by Lloyd Mallan, published by The John Day Co., New York, 1971; NASA, no. MA9-50; Larry Sherer, courtesy USAF, from *Suiting Up for Space: The Evolution of the Space Suit,* by Lloyd Mallan, published by The John Day Co., New York, 1971; NASA, no. 65-HC-549; Larry Sherer, courtesy Litton Industries, Inc., from *Suiting Up for Space: The Evolution of the Space Suit,* by Lloyd Mallan, published by The John Day Co., New York, 1971. 114: International Latex Corp.; National Air and Space Museum Archives; NASA, no. 87-HC-163. 117: NASA, no. 348-26979. 121-123: Art by Yvonne Gensurowsky of Stansbury, Ronsaville and Wood, Inc. 124, 125: Tass; Sovfoto; Tass—Sovfoto; Tass. Hand tinting of black-and-white prints by Tom Kochel, art by Yvonne Gensurowsky of Stansbury, Ronsaville and Wood, Inc. 126, 127: Tass; except bottom right, Sovfoto. Hand tinting of black-and-white prints by Tom Kochel, art by Yvonne Gensurowsky of Stansbury, Ronsaville and Wood, Inc. 128, 129: Sovfoto; Tass—Tass; Sovfoto; Tass. Hand tinting of black-and-white prints by Tom Kochel, art by Yvonne Gensurowsky of Stansbury, Ronsaville and Wood, Inc. 130, 131: Sisson/© Sipa Press. 132, 133: © 1989 Roger Ressmeyer/Starlight.

Time-Life Books Inc.
is a wholly owned subsidiary of
**TIME INCORPORATED**

*Editor-in-Chief:* Jason McManus
*Chairman and Chief Executive Officer:*
J. Richard Munro
*President and Chief Operating Officer:*
N. J. Nicholas, Jr.
*Editorial Director:* Richard B. Stolley

**THE TIME INC. BOOK COMPANY**
*President and Chief Executive Officer:*
Kelso F. Sutton
*President, Time Inc. Books Direct:*
Christopher T. Linen

**TIME-LIFE BOOKS INC.**
EDITOR: George Constable
*Executive Editor:* Ellen Phillips
*Director of Design:* Louis Klein
*Director of Editorial Resources:* Phyllis K. Wise
*Editorial Board:* Russell B. Adams, Jr., Dale M.
Brown, Roberta Conlan, Thomas H. Flaherty, Lee
Hassig, Donia Ann Steele, Rosalind Stubenberg
*Director of Photography and Research:*
John Conrad Weiser
*Assistant Director of Editorial Resources:*
Elise Ritter Gibson

PRESIDENT: John M. Fahey, Jr.
*Senior Vice Presidents:* Robert M. DeSena, James
L. Mercer, Paul R. Stewart, Joseph J. Ward
*Vice Presidents:* Stephen L. Bair, Stephen L.
Goldstein, Juanita T. James, Andrew P. Kaplan,
Carol Kaplan, Susan J. Maruyama, Robert H.
Smith
*Supervisor of Quality Control:* James King

PUBLISHER: Joseph J. Ward

Editorial Operations
*Copy Chief:* Diane Ullius
*Production:* Celia Beattie
*Library:* Louise D. Forstall

*Correspondents:* Elisabeth Kraemer-Singh (Bonn);
Christina Lieberman (New York); Maria Vincenza
Aloisi (Paris); Ann Natanson (Rome). Valuable
assistance was also provided by Judy Aspinall
(London); Christine Hinze (London); Felix
Rosenthal (Moscow); Elizabeth Brown (New York).

**VOYAGE THROUGH THE UNIVERSE**

SERIES DIRECTOR: Roberta Conlan
*Series Administrator:* Judith W. Shanks

Editorial Staff for *Outbound*
*Designer:* Ellen Robling
*Associate Editors:* Tina McDowell, Blaine
Marshall (pictures)
*Text Editors:* Pat Daniels (principal), Peter Pocock
*Researchers:* Karin Kinney, Mary H. McCarthy
*Writer:* Esther Ferington
*Assistant Designer:* Barbara M. Sheppard
*Copy Coordinator:* Darcie Conner Johnston
*Picture Coordinator:* Ruth J. Moss
*Editorial Assistant:* Jayne A. L. Dover

*Special Contributors:* Deborah Blum, Sarah
Brash, James Dawson, Gina Maranto, James
Merritt, Steve Olson, Ron Sauder, Chuck Smith,
M. Mitchell Waldrop, Robert White (text); Sydney
Baily, Vilasini Balakrishnan, Susan Bender,
Andrea Corell, Adam Dennis, Jocelyn Lindsay,
Eugenia S. Scharf, Elizabeth Thompson
(research); Barbara L. Klein (index)

**CONSULTANTS**
RUSSELL BARDOS is the director of shuttle propulsion for NASA Headquarters in Washington, D.C.

BRIAN BIRCH, who has been involved with spacesuit design for twenty years, is deputy program
manager for the extravehicular mobility unit at
United Technologies Corporation, Hamilton Standard Division.

B. J. BLUTH is an expert on Soviet space stations at
NASA's Space Station Program Office, System Engineering Division, Reston, Virginia.

LILLIAN D. KOZLOSKI is a museum specialist on
spacesuits for the Smithsonian Institution's National Air and Space Museum in Washington, D.C.

CHRISTOPHER C. KRAFT, JR., now an aerospace
consultant, was flight director for all Mercury missions and many Gemini missions and served for ten
years as director of NASA's Lyndon B. Johnson
Space Center in Houston.

DAVID LONG is employed at Lyndon B. Johnson
Space Center, where he designs guidance systems
for getting spacecraft into low Earth orbit.

PAUL D. LOWMAN, JR., is a geologist specializing
in comparative planetology. In the 1960s, while at
NASA Headquarters in Washington, D.C., he was
involved in planning a lunar base.

JAMES E. OBERG, an expert on Soviet space activities, is a space-flight operations engineer at Lyndon B. Johnson Space Center.

FREDERICK I. ORDWAY III has worked in rocketry
and astronautics since his involvement with missile
development in Huntsville, Alabama, in the 1950s.
He currently consults on satellite applications.

CHARLES R. REDMOND is public affairs officer in
the Office of Space Science and Applications at
NASA Headquarters in Washington, D.C.

JOEL TAFT is a special-projects manager at Grumman Technical Services in Titusville, Florida.

MICHAEL TIGGES, a senior engineer in mission
planning at Lyndon B. Johnson Space Center, designs guidance systems for vehicle reentry.

FREDRICK W. WEBER, vice president of Contel
Federal Systems, is in charge of the tracking and
data relay satellite systems that Contel owns and
operates for NASA. He has been involved in the
design, manufacture, launch, and mission operations for forty-eight spacecraft.

ERNEST G. WILSON was on-site head of communications systems for Grumman Aerospace Corporation at the John F. Kennedy Space Center during
the Apollo program.

**Library of Congress Cataloging in
Publication Data**
Outbound/by the editors of Time-Life Books.
p. cm. (Voyage through the universe).
Bibliography: p.
Includes index.
ISBN 0-8094-6875-1
ISBN 0-8094-6876-X (lib. bdg.)
1. Astronautics—Popular works. I. Time-Life
Books. II. Series.
TL793.O88 1989
629.4—dc19 88-33928 CIP

For information on and a full description of
any of the Time-Life Books series, please call
1-800-621-7026 or write:
Reader Information
Time-Life Customer Service
P.O. Box C-32068
Richmond, Virginia 23261-2068